P9-CRO-211

THE ART OF
DESIGN MANAGEMENT

THE TIFFANY-WHARTON LECTURES

Walter Hoving

Louis I. Kahn

Sir Misha Black

Van Day Truex

George O'Brien

Edgar Kaufmann, Jr.

Thomas J. Watson, Jr.

Nancy Hanks

ON CORPORATE DESIGN MANAGEMENT

THE ART
OF
DESIGN
MANAGEMENT

DESIGN IN
AMERICAN BUSINESS

TIFFANY & CO.
NEW YORK

Copyright © 1975 by Tiffany & Co.
All rights reserved
Library of Congress Catalogue Card Number: 74-33703
ISBN: 0-8122-7686-8
Printed in the United States of America

Acknowledgments

Special recognition is due several persons who were so helpful in the preparation and editing of this volume. Mr. Richard Saul Wurman for his assistance in reviewing and editing the Louis I. Kahn transcription and essay, *Architecture and Human Agreement*. Ms. Duane Garrison, Director of Publicity, Tiffany & Co., New York, provided the necessary photographs and editorial assistance for the lectures of Walter Hoving and George O'Brien. Ms. Ann Wilson, Managing Editor, *Architecture Plus*, New York, provided the necessary photographs and information for the illustrations to the Louis I. Kahn essay. Special acknowledgement is due Ms. Gay Stebbins, formerly of The Wharton School, for editing the original transcriptions of the lectures.

Contents

Foreword

As one who has long believed that good design is so fundamental to all human endeavor, I am pleased to find a growing concern among executives about the quality of design and aesthetics in American enterprise. The Tiffany-Wharton Lectures on Corporate Design Management are an important expression of this new awareness. More than merely exploring aspects of design in products and services and in physical landscape, they represent a determined effort to help all of us who are in business, as well as future executives, develop a greater awareness for the potential of design throughout our society.

Just as I am constantly inviting members of my own institution to explore new avenues for design, I feel American management needs to promote actively a new kind of sensitivity to design in all of its many aspects. While we cannot expect our managers of today or tomorrow to become design experts, we must find the means to enable them to recognize good design so that they will be in a position to promote a more comprehensive design policy. Whether in computers, jewelry, appliances, or packaged goods, top level corporate management needs to learn how to engage, understand, and work with professional designers if we are to make good design a higher priority in our business firms and in our every day life.

It seems to me that if we are to achieve this mission, schools of management have a special role to play in the education of executives. The partnership between the Wharton School and the arts reflects a healthy and growing effort to broaden the traditional management education programs in business schools.

Speaking on behalf of my colleagues in the business world, I hope the Tiffany-Wharton Lecture Series will serve as an inspiration for business leaders, students, and all of us concerned with promoting the highest standards of design excellence. I hope this book, *The Art of Design Management,* will serve as a stimulus to make all of us more conscious of design in all aspects of our life.

My own experiences in helping to bring aesthetics and good design to Tiffany & Co. have shown me how enthusiastically our organization and our customers respond.

It is a real privilege to support the Wharton School's efforts in bringing greater artistic sensitivity to America's current and aspiring business leaders.

WALTER HOVING
Chairman
Tiffany & Co.

Introduction

Often the question is raised about American business leaders and their sensitivity to the aesthetics of design. When faced with decisions about a new building or product, do managers give consideration to design—to the way things look? Given the traditional programs offered in most schools of management, this question may be more ominous for the business leader than for the designer or artist. Most students prepare for management leadership by studying finance, marketing, labor, human relations, personnel, accounting, insurance, and the like. Even in the most progressive schools of management, design sensitivity or "the way things look" rarely finds its way into the management classroom. Today, however, as business leaders are becoming more sensitive to the ways a corporation projects itself to the public, whether in graphics and advertisements, or in products and services, the subject of design is emerging as an important concern.

The question which has probably occurred to many students of American society is why has it taken so long to bring good design into the management decision-making process? Perhaps one answer is in the narrow focus of management education which traditionally has either ignored design or viewed it as a value concept beyond the legitimate scope of a business education. It is difficult to be concerned with taste and aesthetics in a firm when the only business values and measures that count are those traditional ones optimizing revenues, minimizing expenses, and creating and fulfilling supply and demand requirements. Perhaps the more fundamental truth is that design and aesthetics simply have not been a part of the vocabulary of business leaders. Even when business leaders address themselves to the broader topic of corporate social responsiveness, they view the problem as being consumerism, corporate social auditing, regulation, competitive corporate structures, minority interest responsibility programs, and the physical environment. What is evident today, however, is that good design may be in the throes of being recognized as an important element in good management and corporate social responsiveness.

Why is this so? Perhaps there is a newly recognized unifying force that provides a rationale for the inclusion of good design in business. This unifying force is human need. Everything that is made to serve human needs has to be designed by man; conversely, the only basis for design is serving human needs.[1] In illustrating this point, Chermayeff, in *The Design Necessity*, points to a variety of design examples apparent in many areas of society. Statesmen design public and foreign policy programs. Scientists design scientific experiments. "No city, no highway, no traffic light, no spaceship, no transportation system would or could be brought forth without designers."[2]

Before attempting to apply the Chermayeff analogy to business management, we must first examine the basic purposes of business. Traditionally the business motive has been directed to profit–making and serving the needs of the market. However, for a growing number of enlightened business leaders, management students, and even non-Friedmanites, there is a third purpose of business— namely, serving society and the social environment in which business operates.

If we accept the Chermayeff thesis that everything that is made to serve human needs has to be designed by man and the only basis for design is serving human needs, we can show that management has the responsibility for design in the products and services it produces and markets, for everything that is produced and marketed by business is made to serve man's needs. And, if the only basis for design is serving human needs, it follows that design ought to be part of the very management process by which goods and services are produced and marketed.

---

1. Chermayeff, et. al, *The Design Necessity* (Cambridge, Mass.: The MIT Press, 1973), p. 5.

2. *Ibid.*

It is quite clear, then, that design is compatible with the responsibility of management. Design is a means toward accomplishing the end goals of serving markets and generating profits. Furthermore, design is an element in social responsibility. Good design allows "form to complement performance." The way things look is not irrelevant to the way things work: how they work is how they should look.[3]

The value of design, which is obvious to the artist or designer, also makes good business sense. For example, good design can be justified on the grounds of improving effectiveness in large organizations. Good design saves money and time. Design enhances communication between people and simplifies manufacture. Conversely, the lack of design often causes design by default which often means disorder, confusion, and a lack of congruency.

In the foreword to a recent book on corporate social responsibility, John D. Rockefeller, III, notes there are two levels of corporate responsibility.[4] The first is an obligation to "maintain healthy and viable companies." The second is a "higher level" which is the "determination to look beyond the daily and pressing demands of business to the broader issues which must concern us all." Rockefeller concludes that "it is a fallacy that business can prosper—or, indeed, even exist—without regard to broader social concerns."

> Our society now has an urgent need for the creativity and organizational skill that have been the hallmarks of American business in the past. American managerial genius must adjust rapidly to the new values emerging in our society. Our problems have become so difficult and the currents of change so sweeping, that business leaders cannot afford to hesitate or falter in the face of challenge
>
> I have come to believe in recent years that we are in the middle of a revolution in our society second in importance only to the first American Revolution two hundred years ago. It is a

---

3. *Ibid.*

4. John D. Rockefeller, III, "Foreword," in S. Prakash Sethi, *The Unstable Ground: Corporate Social Responsibility in a Dynamic Society* (Los Angeles: Melville Publishing Company, 1974), pp. viii-ix.

humanistic revolution, a gradual revolution, and it is the dynamics of this revolution that demands a new social awareness on the part of all of us, including American business.

The need for good design in business enterprise is a fundamental human want and one which can be satisfied through both levels of Mr. Rockefeller's corporate responsibility paradigm. Good design and good taste in business can serve to "maintain healthy and viable companies" and can be part of "the determination to look beyond the daily and pressing demands of business to the broader issues which must concern us all." And, based upon the identification of our second "humanistic" American Revolution, good design must be and will be an integral part of that awakening.

Almost alone in his plea for the incorporation of design sensitivity and understanding in American management, Mr. Walter Hoving, Chairman of Tiffany & Co., has been the business Jeremiah of good design for over a decade. Often during his distinguished career, Mr. Hoving has complained that foreign competition is out-designing American business because of major changes in business and management attitudes during the past several decades. In a recent article, he cites, for example, four key developments.[5]

1 *The Growth of Big Business.*

American businessmen have had to become more and more specialized as their companies have grown in size. Corporate officers today must, of necessity, be specialists in finance or operations, in personnel or marketing, and so on. This leaves little room for design education.

2. *Lack of Design Awareness in Large Corporations.*

Little corporate pressure exists to entice the executive to become acquainted with the need for good design. Therefore, few business executives have felt the need for an additional basic education in the elements of design.

3. *Lack of Innovation in Management Education.*

Our educational system has not been responsive to the need for good design in business . . . not even the best schools of business

---

5. Walter Hoving, "Are We Underdeveloped In Design?" *New York Times Magazine* 112 (November 18, 1962): 37.

administration have encouraged graduates to acquire such a background.

4. *Concern with Efficiency.*

The great emphasis that modern management puts on efficiency is a powerful element which militates against the improvement of design in the corporation. For a management executive to question the sanctity of efficiency in modern business takes great courage because top management has been taught to literally worship at the shrine of efficiency.

With these observations in mind, Mr. Hoving became convinced that if a new corporate conscience towards the role and implementation of good design and aesthetics were to be generated within the American business community, it had to begin with a leading business school. A plan was then developed which Mr. Hoving felt would enable business school students to become more sensitive to the nature of design and appreciative of its role in American business. Thus, he came to Wharton to explore its interest in his innovative venture. The result was a series of seven lectures on the campus of the University of Pennsylvania sponsored by The Wharton School and Tiffany & Co. These lectures, the first of their kind in an American business school, were designed to expose students to a variety of viewpoints about the nature and application of design as well as provide an incentive for making good design an integral part of management in every kind of institution, from the corporate headquarters, to government offices, to schools, shops, and so on.

Our book, *The Art of Design Management,* was inspired to share these insights with other students, faculty and business persons interested in design. Its organization follows the sequence of the Tiffany Lectures. Walter Hoving's lecture, "The Crisis of Design and Aesthetics in American Management," provides the rationale for the lecture series and gives an account of the bankruptcy of American business in terms of design deficiency.

In the first lecture, George O'Brien, designer for Tiffany & Co., outlines the key responsibility of the design director as an "editor" of good design and professional aesthetician in the coroprate organization.

Louis I. Kahn, in his "Architecture and Human Agreement," provides a poetic explanation of art as the only language of man and as an instrument for revealing the "humanness of man." The exchange between the executive and the designer is a powerful and human experience—the artist is the expressor and the executive is a wanter.

In the third lecture, "A Profitable Art," Edgar Kaufmann, Jr., applies a marketing approach to the determinants of good design and good taste. Kaufmann believes it is the marketplace—consumers, purchasers, and the public—that should decide taste and what it is the designers design.

Sir Misha Black, in "The Designer and the Manager Syndrome," outlines a distinct set of responsibilities for the designer and the manager in order to generate good design quality in business. Sir Misha believes designers and managers should share a responsibility for our environment as well as study the social symbolism which products provide for consumers.

Thomas J. Watson, Jr., in his lecture, "Good Design is Good Business," makes it quite clear how the experiences of Olivetti, Eliot Noyes and a new corporate policy provided the subsequent design and business success for IBM corporation.

In the sixth lecture, "The Environment for Creating Good Design," Van Day Truex identifies the "training of the eye" as being the most important way to gain an appreciation and understanding of good design in business management. Management must abandon its primary orientation of "will it sell?" to ask the more basic question: "how is the product to look?"

Nancy Hanks' concluding lecture on "Design for America's Third Century" emphasizes the important role of the private sector in creating good design in the next century. She notes with optimism the tremendous growth of the National Endowment for the Arts as a positive reflection of a new national concern for design.

Though it may be impossible to gauge the impact of the Tiffany-Wharton Lectures on Corporate Design Management, we are excited about the discussion they generated within The Wharton School and University community. We have published these lectures with the hope that this book will serve as an inspiration for business leaders and students to form a partnership with designers and artisans to help bring into being a new awareness of good design to corporate America. And, for reasons beyond economic necessity, is it not unreasonable to forecast that design quality and sensitivity will become a part of corporate America's social consciousness?

THOMAS F. SCHUTTE
February, 1975

The Crisis of Design and Aesthetics in American Management

WALTER HOVING
Chairman of the Board, Tiffany & Co.

and

GEORGE O'BRIEN
Vice-President and Design Director, Tiffany & Co.

Walter Hoving

There is an abundance of good designers in America. The basic problem is that corporate management does not know what good design is and would not know what to do with a good designer if it had one. Design management consists of establishing corporate design policy, organizing design policy and procedure within the firm, and determining and evaluating the design concept and program of the firm. Business schools in America have done nothing about corporate management's responsibility for good design and aesthetics in business. Producers and retailers are often preoccupied with a concern for "Will it sell?" Most people walk in ugliness in terms of architecture and merchandise. American business firms ought to have a top ranking corporate aesthetician who manages the design concept, standards, policies, and programs of the firm.

1

We have plenty of very good designers in America. This is due primarily to the number of excellent design schools found in many parts of the country. We do not have, however, enough corporate executives who understand what good design is or what to do with a designer when they see one. It is for this reason that I address myself to those who are soon to enter the corporate structure. By corporate design management, we mean how to set a design policy within a company, how to organize to carry out that design policy, and how to determine the design concept of the company. At The Wharton School and at other leading business schools, they teach marketing, finance, management, advertising, accounting and so on, and they do it extremely well. But neither Wharton nor any other business school in the country has done anything about the corporate manager's responsibility in the field of design.

Now there are three aspects of all consumer goods that interest people, be it a jewel or an automobile, or anything else. First is the material which must be authentically represented. This is not too difficult to test because there are definite standards. Fourteen karat gold is, for example, fourteen karat gold by law. There is no way of falsifying what it is. The second is the quality of workmanship. This gets into some gray areas because there is difficulty in obtaining an absolute standard. And the third is quality of design which, of course, is an even grayer area because there are no standards that can be absolute. We feel that any merchant, department store executive or manufacturer, in pure honesty, must honor these three concerns. When somebody comes into a store to buy something, they must feel confident that the material is what has been represented. Also, they should feel that the company has been honest about the workmanship. The automobile must run properly or the jewel must be worked with skill. Also, it is important that the company be able to represent its design as being up to the company's standards. In other words, if I try to sell you a glass and I think it looks terrible, I'm dishonest if I tell you it is beautiful. Unfortunately, this is done all the time. Advertisements talk about how beautiful something is when, in fact, the object is dreary looking.

1. Ruby and diamond flower and cone necklace designed by Jean Sahlumberger of Tiffany.
(Courtesy of Tiffany & Co.)

Generally speaking, the public's taste is much better than the storekeeper's taste or the manufacturer's taste. There are several reasons for this. One is that the executive has no taste of his own, or the only thing he is interested in is "Will it sell?" The clerks who then have to sell the item don't like it, which in turn is one reason why there is so much rudeness on the part of salespeople in department stores.

Something should be done about this situation if only for the reason that the whole world is outdesigning America. Italy, France, Germany, England, in fact, every country in Europe outdesigns us. This goes for Japan, too. The question is, why is this so? One reason concerns the actual physical environment within which we live. When you are in Europe, you literally walk in beauty. If you were brought up in Europe, you have lived all your life in Paris, or in Rome, or in Florence, or in London. You have been constantly exposed to beauty. We don't have that same opportunity in this country. Most of our people walk in ugliness. Ugliness in architecture. Ugliness in merchandise. In the old days, one used to say that what made a thing good was because it was imported. We don't say that anymore. We have taught the rest of the world how to mass produce and how to mass market. The fact that they have always outdesigned us makes it very dangerous for business in this country.

If business in this country cannot remain prosperous, then we know that people will not have jobs, universities will not get the money necessary to function and so on. That is one key reason why good design is so important.

The question is, how does one create a high standard of design? In the first place, the top executive, or one very close to the top executive, must start by intellectualizing company policy as far as design is concerned. That is what most junior executives are going to confront when they move upwards in the corporate world. What are they going to do when the question of design comes up? When I have attended design meetings, I have seen executives who did not have the slightest concept of what design meant. What is worse, executives who would not dare to give an opinion about law because they were not trained as lawyers would glibly respond to questions about design on a purely subjective, spontaneous level saying—"Don't like it" or "It's pretty." Sometimes they would bring in their wives to give an opinion. This is even worse because the Chairman's wife feels she outranks the President's wife, and the Vice Presidents' wives, and they will naturally parrot her views. So, my first point is that you have to have an intellectual conception of design policy. That means that you have to know something about aesthetics in order to think fundamentally about design, which is a hard thing to do.

It is my feeling that trained designers will become excellent designers, if they get the opportunity. They will go from door to door, from company to company to get this chance. The trouble is that the executives often do not know what to do with them, so you find many designers are taking jobs elsewhere because they cannot utilize their training in design. But you cannot leave questions on design policy to designers anymore than you can leave war to generals. There has to be somebody over the designers who is an aesthetician, who knows the company's design policy and who can guide designers. This is what we have done at Tiffany's. From my own experience, I would say that design presents a more difficult problem even than marketing, advertising, merchandising or finance, because those things lend themselves to a more accurate measure. They are accepted practices in company after company,

2. Place Vendome, Paris.
   (Courtesy of Air France)

but very few companies have concerned themselves with design and aesthetics. We hope that schools of business administration will become concerned with aesthetics as it relates to corporate decision-making whether in terms of product design, architecture or interior design, and we strongly believe this kind of intellectual preparation should be incorporated into the graduate programs for people who are going to become the leaders of the business world thirty years from now.

George O'Brien

> *Design is as much a part of communication as the written word. In America, "the general level of design is abysmal and caters to the lowest denominator, or even taste." A design director must practice objectivity of taste and must ask key questions about design on the basis of "Is the design good?" "Is there a valid expression of a decorative style?" and "Does the design enable the product to function for the user?" The ultimate in the management good design is the "common bond in the quality of design, manufacture, and material."*

Strictly speaking, design in the visual and applied arts is an arrangement of certain elements, including line, form and color, to create and express an idea or an emotion. It is as much a part of communication as the written word. Design is the visual basis of the industrial and decorative arts, as well as the fine arts, painting, and sculpture. I am primarily concerned with industrial design or the application of design to useful decorative objects and spaces. This does not mean, however, that the fine arts are to be ignored, for an appreciation of them is essential to the understanding of industrial design. They share a vocabulary and a common goal, namely the visual solution of an idea. In fine arts, the idea is totally intellectual and involves emotions. In industrial arts, the idea is more concerned with the actual purpose of the object being designed. Granted, people can get quite emotional about designed objects such as automobiles, but the purpose of a car is primarily to provide a useful machine that should function well and look well. This is not always a successful marriage, as is shown by what is coming out of Detroit today. In the nineteenth century, architect Louis Sullivan, the father of the skyscraper, decreed that "form follows function." This has become the rallying cry of modern design. Detroit, unfortunately, was not listening.

About design awareness: some people have an aptitude for it, just as one has an aptitude for science or sports. Others have to work harder to achieve it, but it is not an impossible task and the pleasure it affords is well worth the effort. There are many books dealing with design, and reading them is essential. But the real work begins when we start to educate the eye. This is done by visiting museums. Keep your eyes open all day, every day, everywhere. Look at nature, paintings, sculptures, buildings, furniture, clothes, people. Be a camera and take in the visual. Then digest what you see and attempt to determine why one design is successful and another fails. All this is work, but well worth it. One way to judge a good design is to look at the object in question as you would a person. Is it interesting and exciting? Has it character and personality? Is it honest and sincere? Or is it just insipid, dull, and boring? Always be objective: save the subjectivity for the time when the eye and intellect have realized the basic education. Only then can one specialize.

As a Design Director at Tiffany's, my main responsibility is the product. In addition to jewelry, Tiffany and Company sells silver, china, crystal, clocks and stationery, all areas in which I, as a Design Director, work. We design and make in our own factory 85% of the silver we carry. The balance is designed by us but made elsewhere, mostly in Europe. This is so because the design skills have dried up in America. In other categories, we do not manufacture the glass and china or the paper but we do design over 75% of our stock. The balance is selected from manufacturers around the world with a regrettably small showing of American manufacturers. While American produced goods are well made, the general level of the design is abysmal and caters to the lowest denominator of design, or even taste.

My purpose is to, hopefully, contribute to the improvement of this standard of design and to encourage good design in American products. A Design Director is, in effect, really an editor. While I do designs myself, my real task is editing the work of designers, encouraging them, bringing them back to the intent of a design if they wander too far afield in fancy, and trying to do all this without breaking their design spirit. In working with manufacturers, I try to convey the Tiffany point of view effectively enough for them to understand what it is we are trying to achieve. And the Tiffany point of view is basically one of respect for quality in design, manufacture, and material.

Our sense of design is such that it is generally understated and avoids pretension. This is not to say design is without a sense of humor or that we are not concerned with performance. We work also to achieve a quality of timelessness in our design. We don't like to think that what something bought today in the store will in ten years be regarded a "1973 fad." For example, while we do not have anything that is pure art deco, we do have things that echo the best of art deco. And this is what we mean when we say we try to have a "timelessness" in our design. Furthermore, the costliness of a design does not affect the amount of care we spend on any given object. We are just as painstaking about a $2.00 earthenware dessert plate as we are about a $2,000.00 silver bowl. To our way of thinking, "a point of view" is about the most important aspect of any undertaking. It creates a cohesion in a collection of designs that cover many uses and many decorative styles.

When deciding what we should present to customers, the Design Director must constantly practice an objectivity of taste. It is not a question of whether or not I like it *personally*, but rather, do I like it *professionally?* Is it a good design? Is it a valid expression of a decorative style? Does it work ? We have many customers with many preferences. Some like modern, some prefer traditional, some like art deco. The Design Director's job is to see that they have the best of all design periods offered to them. Although it is generally true that good design from any period will mix together with that of another, this works only if all the designs are good. In display, it is easy to separate designs, to put modern in one place and traditional in another, but it is much more difficult to put them all together without a visual quarrel. At Tiffany's we are against design segregation. All of our designs are displayed by purpose of use, not by

3. Tiffany Glassware.
   (Courtesy of Tiffany & Co.)

period or style. So it is not unlikely to find in our China Department a cream jug, originally designed by Josiah Wedgwood in the 18th Century, living happily next to a sugar bowl made last year in Japan. They both share the Tiffany point of view—common bond in the quality of design, manufacture, and material.

The presentation of our design is an equally important aspect of the impact of our goods. All the interiors of our stores are designed to be an effective background for the design. The furnishings must not intrude; they should enhance. The same hold for sales and customer education, as well as advertising. Before I venture into some specific Tiffany designs, I would like to quote a very successful manufacturer who played good design into a world-wide business— Ernst Braun, the founder of the Braun Electric Co. in Germany: "We like to think of our products as the perfect English butler—they never intrude, but they are always there if they are needed."

In attempting to apply theory to the actual design problems we have at Tiffany, I want to first take a basic design idea such as the ancient Chinese and Japanese vase shape. One of our collections of crystal began when I was working in Japan with some manufactur- ers who were doing things of Scandinavian origin (Fig. 3). They

4. Drab Ware china, manufactured for Tiffany by Wedgwood.
   (Courtesy of Tiffany & Co.)

didn't have the Scandinavian feel for glass, and their products looked pretty lifeless. Crystal making is not a time honored tradition in Japan—only since the late 19th century have they been doing this. So the problem was how could we take advantage of the skills of the Japanese workmen, but come up with a product that would also reflect their design heritage, something that they would understand. It occurred to me that nothing is as beautiful and timeless as the oriental vase shape. We then gathered together a balanced selection of vase shapes, made scale drawings, and the result can be seen in our current collection. Though the design source is ancient, the interpretation in crystal makes the vases absolutely new.

Another example of reaching into the past for something that is very modern is our Drab Ware china which is made for us by Wedgewood in England (Fig. 4). The shapes of the china were devised by Josiah Wedgewood in the mid-18th century, and the color was introduced by Wedgewood in the mid-19th century. We found the shapes and color in the Wedgewood Museum and begged them to produce it for us, as it looked absolutely right for today— and it proved to be so. We have sold more of the Drab Ware china in

5. The abstract look of this fine earthenware Chinese bowl seems very modern— it is actually 900 years old. (Courtesy of Tiffany & Co.)

four years than we have of any of our other patterns of any kind of china in twenty-five years. It is a good example of timelessness.

We have also taken a commonplace object and raised it to a work of art. Take, for example, the classic French stockpot, found in every French restaurant and practically every French home (Fig. 7). It always sits on the back of the stove as you add things to it from time to time, and it is always simmering away with its stock, the base of all sauces. The size that we translated it into, in sterling silver,

6. The same earthenware design was translated into silver by Tiffany—yet another example of employing classic design for modern products. (Courtesy of Tiffany & Co.)

7. Sterling silver stockpot—commonplace, but classic and functional. (Courtesy of Tiffany & Co.)

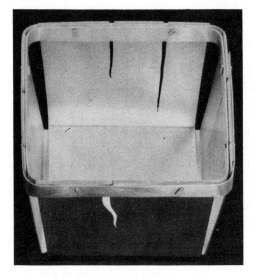

8. Sterling silver strawberry basket. (Courtesy of Tiffany & Co.)

9. Japanese porcelains.
   (Courtesy of Tiffany & Co.)

is just exactly the same as the original copper pot. You might ask the question—Why do a stockpot in silver? The answer is, simply, that we like it. First of all, it is a beautiful design. It is classic and perfect. Secondly, it is useful. It can hold flowers, or soup or be used as an ice bucket. And the thing we also like about it, its added bonus, is that it is amusing and we like this humorous quality.

This same idea is exhibited in our sterling silver strawberry basket

10. American spongeware china, manufactured for Tiffany by Stangle Pottery.
    (Courtesy of Tiffany & Co.)

(Fig. 8). It has been translated with such degree of accuracy that we have even included the rivets that hold the thin wood together and the actual splits in the wood. This is intended as a decorative object but it can hold anything.

Part of our collection of Japanese porcelains which are made today originate from classic Japanese porcelain (Fig. 9). But the painting which expresses the feeling of Imari, a school of Japanese decoration, is a new one. It is new in that its origin came from a Nō drama costume which we interpreted for porcelain. It is all hand painted and each piece is hand thrown in Japan exclusively for us. One of our two American representations in china is made for us by Stangl Pottery in New Jersey (Fig. 10). Its origin was the early 19th Century American pattern called spongeware. The technique they used was to take a sponge and dip it into paint and put it all over the

11. Crystal paperweights
(alphabet blocks).
(Courtesy of
Tiffany & Co.)

object. Then they would fire it to give a color and a decorative look. The best known original spongeware of the 19th Century was blue and white. We started off with blue and then we went to green and then to yellow. We have been working on this for about five years and every year we add some more pieces. This is, you know, our heritage. The American heritage. It's perfect for today. It is useful; it is dishwasher proof; it is even oven proof. Good design should not only be good looking but practical.

We are always looking for something that would make an amusing and fairly inexpensive gift. We have, for instance, a paper weight in the shape of an alphabet block, which we decided to do in crystal (Fig. 11). Getting a full lead crystal cube, two inches square

12. Part of the Tiffany stationery collection.
                    (Courtesy of Tiffany & Co.)

with an initial, down to $10 in price was really quite an achievement.

I finally want to include our stationary as an example of the selection that we offer people (Fig. 12). Knowing that there are many different kinds of customers with many preferences, we try to balance the presentation of our products. Some stationary is a little dramatic, some is classic, some has more vivacity with bold color but it still sits right next to the staid, the classic and the traditional patterns. We believe in this kind of diversity at Tiffany.

Architecture and Human Agreement

Louis I. Kahn,
Architect

> *Art is really the only language of man. Man lives to express. Form is the realization of that which can exist or rather which can remain present. Presence is what design is all about. Direction for the expression of a product can only come from an individual—not a committee. Only after man's expression of faith in what is realized can there be a fruitful exchange which can make the executive and designer better for each other and better realizers. The executive and designer are like gamblers, because to achieve the overwhelming desire to see something "which is worthy come forward" requires a sense of freedom to do things other people wouldn't dare do.*

I want to make just a few distinctions because I live by them in the work I do, and I think they are of a general nature and part of anything that a man does.

Essentially, every man deals with art.

Even the way a man walks, the way he uses his language, the way he turns a page, or his conduct in a restaurant has to do with some form of art. It is choice. It is noticing the grace of your actions.

Art is really the only language of man.

All other languages serve to bring art into the fore.

17

Man lives to express.

The reason for living is to express.

What comes to my mind is the first feeling of man when his eyes were opened and he saw things around him. I don't mean a "certain time" when man was born. I cannot conceive of man before human. Human seems to be that which emerged out of being that species, man. And human is what we really talk about and not necessarily the species we live by.

The first sense must have been beauty.

Beauty as total harmony.

Not the beautiful, not the very beautiful, because they are already less than beauty itself in the same way as *very good* is less than *good*. There is something about the unmeasurable word which when made measurable becomes less.

Art strives to communicate in a way which reveals the human. It is what yet has not been that you're dealing with. You are not dealing with something that is recognizable before it is made. Even as it is described, you don't know what it is until it is interpreted.

Then, beauty is the first feeling.

The feeling of some over-all harmony that is accepted without any doubt whatsoever. It is only knowledge that comes in which helps to somewhat modify the beauty of not knowing, and the realization that there is a tremendous rapport between what is evident and what is yet not expressed.

Possibly the first counter feeling or sympathetic feeling is that of wonder.

Wonder again, free of knowledge, free of knowing. Just a sense of being in the presence that motivates an overwhelming feeling of wonder. From it, almost on its heels, is a realization that it must be so. You're in touch with the universe, not with just knowledge about laws which are only touching the content of order, of all laws: those that are made, those that are invented, and those that are still not invented. From realization comes an all-important working force which we know as form.

Form is the realization of that which can exist, or rather, which can remain present.

When you feel form, you feel the inseparable parts of something.

When you take one part away from form, all of it disappears; it cannot hold together. Form is a realization of that which can be, and you can already sense the inseparable parts. There is no question of its nature.

Form is completely different from shape.

Shape is something that a man chooses to interpret form.

Existence is not presence.

Presence is what design is all about—to give presence to something and to translate the realizations which come from the sense of form; to preserve that sense with everything that you have at your disposal and to give presence to that which has existence in the mind.

When an executive feels the need for the true expression of his product, it must have in it also the executive's faith in his product, and that he knows in no way is the product phony. It has a right to be given presence. That message, and its inherent motivation, the executive can convey to the artist or to the designer.

It must be recognized that every designer is not like another designer. There is really a question of singularity to singularity. If you get a direction from a committee, I am positive the product will be less, the expression will be less. If it can be in an individual, it will have such resources that a committee meeting many, many times would never have. The individual has the ability to see it all as a unit. From that sketchy first realization mixed with faith in what is realized, can there be exchange of a designer and the man who wants the design made. There can be a fruitful exchange which can make the executive a better executive and the designer a better designer.

It's humans, human, and a human.

The artist is an expressor, and you might say the executive is a wanter or he is a desirer, you see, and both are powerful together.

If the designer is a specialist in certain kinds of designs, he is not a designer at all. He becomes a hack, and that's all. The designer is only important when he reaches a subject absolutely freshly as though he never knew anything like it before. His whole desire, really his only desire, is to make that presence which is yet not present.

That is the very essence of desire, that which is not yet said, yet not made.

It has nothing to do with need.

If you need a designer, then you're actually wanting less than if you desire one

Singularity to singularity is the very act of teacher and student.

I think that's what teaching is all about, and instruction is something else.

I think really the executive must be the teacher because in him lies some kind of secret as to what he wants to attain which the designer can give him. I would see even such an instrument, born out of nature, as a locomotive—I can assure you that one man doing the same instrument with the same premises in back of it would design it differently than another. That's how delicate is the choice of a designer.

There is something I must talk about because it possesses my mind, I simply must. It is the position of every individual—every individual being completely different from the other. I mentioned the desire to express as being an aura which sits within us, all of us, in some form or another. The farmer has it. The architect has it. The businessman has it. The painter has it. And the other powerful force is nature itself, which I say is the maker of all presences. What is desired can only be made present by knowing the laws of nature or better still, the order itself.

There is only one law really—order.

In order, there is no chaos.

Chaos only exists in man's mind, but not in nature.

Nature is non-conscious; it cannot be chaotic.

The travel from the desire to express to the means to express, which is all nature, varies in each individual. The scientist would hold back his sense of desire and allow nature to come to it. By holding it back, he holds back the movement of what may be any poetic thought to avoid that which may make what he is searching for have too much singularity, or better still, too much of the undefineable or unmeasurable limits because he is looking for the measurable. And he looks for every means to get the measurable pure of what may be his other senses, and that is of desire itself.

When the poet marches toward the means and marches for a long, long time, hoping the means is never necessary—but he finds the means is necessary because he has to print things—he has to say something; you can't just leave out all words, which is his real intention. And so nature is held back, that is, the maker, the means; and he is able to send forth that which has the greatest power of transcendence in his poetry.

Everyone has this balance of the measurable and unmeasurable in him; but a man like Einstein would be one like the poet who resists knowledge because he knows that if he were to talk about what he knows, he knows that it is only a miniscule part of what is yet to be known. Therefore, he does not trust knowledge but looks for that which can be order itself. He travels like the poet, the great distance, resisting knowledge; and then when he does get a smidgen of knowledge, already he reconstructs the universe.

But the executive, if he truly has in him the aura or the brilliance of what is a kind of totality of his purposes, becomes equally as intangible as the artist who also deals with the intangible at first before he makes it tangible. The executive must know what he wants, and he must turn to someone who knows how to express it. They are both gamblers in a sense, great gamblers, because that sense, the desire to see something which is worthy come forward, is an overwhelming feeling and there is nothing really that is respected for merely the operational. It is something very much like the daring acts of the first people in the United States, who by their sense of freedom did things which other people wouldn't dare to do.

A man in business or in the professions feels, like a flash, the aura of the inseparable aspects of the work that he is engaged in. He knows that if he tries to incorporate bits of other things, that he goes away from the power of what he is trying to put into being or trying to express. The designer takes what is visualized as the inseparable parts, takes these inseparable parts or elements, and applies them to what nature can make. He will not make a design which is just there for the looks of it. It must be something that answers the laws of nature and allows it to take a shape true to the elements. In early industrial design, it was nothing to make a streamlined comb that for all the world looked like something that is to be hurled into space, as though it were made for the purpose of movement, just to give it a nice shape. That was true of egg beaters, toasters, and many other things. It was also made to be applicable to a car that mostly goes through town at a speed less powerful than what it is able to employ.

An airplane is made without those considerations of selling. They are made really for the very strict laws of nature. They take a shape that is very true to themselves, and the expressions of it are not yet finished. It's not finished for one reason—the desire in man will never be satisfied. I think that if planes were to take but two hours to get to California, I think nobody would raise an eyebrow. But if you were to touch the desires of man, you'll find that what man really wants is a flying carpet which would take him anywhere for no cost at all. He would need no pilot, not need any gas and he would really have it under his bed and do what he wants with it. I am sure this desire will never be realized, at least not in my time.

But the nature of desire is insatiable. This motivation will always exist where there is any possibility of creating a sense of the completeness in something before it actually has presence. This "will to express" is an intuitive sense which is after all only the sense of how we were made.

All of learning is answerable only to the desire to know how we were made.

It is also the reason for living.

I think that there are no formulas which would make a good choice of a good designer. Rather it stems from the man who has faith and the sense of a certain completeness which he might not be able to describe but knows through his intuitive sense, and through what may only be a smidgen of knowledge, that it can be. The designer can make it because the undeniable force of faith in what you are doing is contagious. The desire is there, and the artist can pick it out and move with it and do more with that than the most minute instructions of how it should be. He is anxious to test his ability to make what is yet not, into being.

When I speak about silence and light or the desire to express and the means, I say that all material is spent light. Light that has become exhausted. Creation makes me think of two brothers who were really not two brothers. One had the desire to be, to express; the other had the desire to be something that becomes tangible, something which makes the instrument upon which the spirit of man can express itself. If the will to be is to become something of the predominance or the prevalence of the luminous, then the luminous will turn into a wild dance of flame, spending itself into material. And this material, this little lump, this crumpled lump, made the mountains, the streams, the atmosphere—and ourselves.

We come from spent light.

I see a drawing of the section around Siena by Leonardo da Vinci. Doesn't it look like a crumpled kind of thing which looks like spent light? He's such a marvelous artist. I'm sure he didn't think of this particular theory, but in him I'm positive that he had sense of wonder enough to make such a drawing. I'm sure it's an inaccurate map; but one that's better than a map. He tells you somehow that this is the product of light.

Light is an important part of architecture, and the way I go about learning about it is to make certain assumptions. And when I learned about Greek architecture, it never occurred to me that the column is where the light is not and the space between is where the light is. So it is a march of no light, light, no light, light and it grows out of the wall. It grows out of the parting of the wall into the column. And that's the marvel of the artist.

Isn't it a miracle that a man, with no examples around him, can make such a thing manifest as a column which grows out of the wall and makes its own window by the rhythm of the lightless, light, lightless, light?

In drawing, it's the same thing. You can say that drawing is made by the stroke of a pen where the light was not.

Look at a drawing.

Look at the sources of light.

How it marches with the light, and then disappears where the light cannot reach, into places where it goes into reverse because the direction of light is lost.

The stroke of the pen is where the light is not. Now you can teach drawing that way. Of course, there are other ways too.

Here is the religion of light which I think an architect must have. Every building, every room must be in natural light because natural

1. Alfred Newton Richards Medical Research Building, University of Pennsylvania, Philadelphia. Architect, Louis I. Kahn.
"It can be said that the sun never knew how great it was until it struck the side of a building." (p. 25)
(Robert C. Lautman)

2. Interior, Kimbell Art Museum, Fort Worth, Texas.
Architect, Louis I. Kahn.
"Windows cause glare; so windows were not considered. But light from above, which is the most brilliant, was considered the only acceptable light."—L. Kahn[1]
(Marshall D. Meyers)

light gives you the mood of the day; the season of the year is brought into your room.

It can be said that the sun never knew how great it was until it struck the side of a building.

And so we can also say that when a light enters a room, it is your light and nobody else's. In the Kimbell Art Museum, I used a natural light fixture. The light comes from on top, and Richard Kelly, after some kind of intuitive design I made, put the thing in a computer and came out with the shape which sent the light along the cycloid vaults and from which the light is gotten for all the paintings (Fig. 2). This light is less injurious than artificial light. The argument that a

1. "The Mind of Louis Kahn," *The Architectural Forum* 137 (July/August 1972): 60.

3. Exterior, Kimbell Art Museum, Fort Worth, Texas. Architect, Louis I. Kahn.
Vaulted areas (cycloids) are joined by clearly defined connectors that at once isolate and join.
(Marshall D. Meyers)

museum should not have any natural light and have just walls can be questioned.

Because the vaults can span 100 feet without a column, I expressed the ability of the vault to be a beam, and didn't allow anything that enclosed the building, any wall, to touch this characteristic of construction. (Fig. 3).

The first plan I made for Dacca, Bangladesh, then the second capital of Pakistan, started with just the slightest hint. I saw the devotion of the Moslems to their prayers five times a day, and I was inspired very much by the anonymity—that's probably the wrong word—but that the mosque need not be visited. It was there for those who wanted to go. There was no preaching there. You simply said your prayers. It was just a community building which was your community building, nobody conducting it. And I thought that it should be part of the Assembly—that the Assembly should look to the mosque, and the mosque should look to the Assembly. It was a

4. Residences for legislators, Dacca, Bangladesh. Architect, Louis I. Kahn. (Anwar Hosain, courtesy of *The Architectural Forum*)

5. Arcade, National Hospital, Dacca, Bangladesh. Architect, Louis I. Kahn.
"The outside is one building that belongs to the sun. The interior belongs to
the shadows, or the place where people live." (p. 29)
    (Anwar Hosain, courtesy of *The Architectural Forum*)

religious state and that's what made me think of it. And so this was
the citadel of the Assembly, around which the members of the
Assembly—the judges and other dignitaries of the Assembly—were
to live on a lake (Fig. 4). Across it was the citadel of the
institutions—symbolic institutions—because I felt that legislation is
that which brings about the institutions of man or which makes
things available to the people and expressive of a way of life. I
began it that way, and it remains this way.

One feature which was very important to me was to make the
distinction between the place of legislation and the supreme court.

Legislation is circumstantial law and the supreme court is law in
relation to humans.

When I presented this idea to the Chief Justice of the Supreme
Court, he said that he didn't want to be near the Assembly. And

6. Hurva Synagogue, Jerusalem. Preliminary sketch by Louis I. Kahn.
"I sensed the light of a candle plays an important part in Judaism. The
pylons belong to the candle service and have niches facing the chamber. I
felt this was an extension of the source of religion as well as an extension of
the practice of Judaism."—L. Kahn[2]
(Courtesy of *The Architectural Forum*)

when I sketched for him the mosque which was there, he took back
his words and said, "The mosque is sufficient insulation for me." So
that was the basis of the plan. The expressor worked closely with
that which had to be expressed—the orders of the government and
the man who was entrusted with this expression. Mind you, somebody else would not do it the same way.

There are great porches behind the windows (Fig. 5). You use no
windows here because the heat is very great. The sun is very strong
and all windows must recede back to where the cool air can open to
the rooms. The outside is one building that belongs to the sun. The

2. *Ibid.*, p. 69.

interior belongs to the shadows, or the place where people live. Now these were not done with blinds and other accoutrements. They were simply made architectural by expressing fully the powers of architecture.

Again expressive of the same idea is the Hurva Synagogue in Jerusalem, where I'm using the stone of the Western Wall and inside of it is a concrete structure (Fig. 6). These stones are cool compared to any other structure. The concrete is warm because the reinforcing rods get heated up—it's a warm construction.

When Jonas Salk came to me, he said he wanted to have a laboratory to which he could invite Picasso. Well, when he said that, I came up with the idea that what he really wanted was a place of the measurable, which is the laboratory, and a place of the unmeasurable, which would be the meeting place. Biology is not just a scientific or a simple task of finding which is measurable. Salk realized that even a microbe wants to be a microbe, which has something to do with desire. There is this unmeasurable quality even in matters scientific. So I, as the interpreter, gave him a plan that was terribly expensive which was why it was not built. But this design concept did motivate the making of what was built, because we reserved space not for what would ordinarily be stuffed in this laboratory building, but for the meetinghouse—the art library, the gymnasium, the dining room.

Is there a better seminar than the dining room?

We worked collaboratively all the time. And I think he is just as much a party to this design as I am. I can say that this was a true collaboration.

Collaboration in the arts is not possible.

But collaboration in that which motivates the arts is possible.

A Profitable Art

Edgar Kaufmann, Jr.
Adjunct Professor of Architecture, Columbia University

"The greatest difficulty in the world of design is an extraordinary blindness about its raw material—the consumer." There is no such thing as the right design for all. Designers and design users have failed because of their neglect in studying the needs and wants of the consumer or the market place. "The market is the mistress and always has been" in design. There is no such thing as universal "good taste." Good taste is a function of what it is the audience, purchasers, and public want.

In dealing with the problems of trying to have an effective attitude toward design in our world today, I feel the first thing I should discuss is "good taste." I am afraid that I do not believe there is such a thing. I think good taste is contingent upon the social situation, the time and the spot, and the people who are involved. And I think the attitude, that there is a particular thing which can be designated as "good taste," is a leftover from an aristocratic society. Fundamentally, it stems from the fact that Louis XIV could hire Mr. Le Brun and introduce an academic framework within which it was possible for a few highly skilled, very talented, highly intelligent people to agree on certain norms of visual expression in which all the arts served the king. And necessarily, since Louis was an absolute

31

monarch, this meant applying these norms in work done for his courtiers. The taste of the court filtered down through the community and eventually showed up even in folk art, though it often took a hundred years to reach that level. This situation was eminently understandable and suitable for such a society; today, however, we do not have this kind of society. There is no one who can enforce standards of taste or even reach any agreement on what constitutes "good taste." The best that one can do is to say, "I have spent a lot of time thinking, looking, experiencing and I come out *here*, but I know from experience, that somebody else with just as much patience, just as much background, and just as much innate good sense may come out somewhere else."

So what does that leave us? Does it leave us with something we might call "pop" taste? Is it really true that, if we go to Las Vegas or if we just go downtown to the nearest neon-lighted corner and see all the jazz and the glitter and the excitement and the confusion, the vulgarity, the brashness, and the really highly entertaining vitality of the lowest level of commercial art, this is our alternative? Is this the good taste of an egalitarian society just as Mr. LeBrun's was the good taste of an authoritarian society? It seems a little hard to accept. It is not impossible, yet I have an idea there is something quite different going on. I think that what is happening, what has been happening for indeed a very long time, is that in a more or less egalitarian society there is an enormous diversity of tastes. There are suitable tastes for different segments of the population, tastes that fit an occasion, a way of life, certain ambitions. But, there are other tastes, quite different and equally valid.

If this completely open idea of taste has any validity, we really come to grips with a third problem (after good taste and pop taste) which is, what in the world does this diversity of taste really look like? Can we spell it out? Can we pin it down? Or can we pin them down? It gets us into the very very tricky realm of statistics. Does it mean anything if you say something has sold well? Or what does it mean if something sells well in New England but it bombs in Texas? What is the real function of analyzing market reactions to taste? This is a very difficult question. People have never yet come up with a final answer. Yet in trying to come up with such an answer lies the hope of having a grip on the design attitudes (plural) which might serve the kind of world we live in and are going to live in for a very extended future.

When considering the analysis of statistics, I cannot neglect Sir Karl Popper, who has a most extraordinary, challenging mind. Born in Germany, he went to live in England, becoming highly esteemed as a philosopher and logician. When I was recently in Britain on a jamboree at the University of Edinburg, I asked a philosopher, a young fellow, eminent but rather pompous, about the state of Popper's reputation in his world these days, that is, at Oxford and Cambridge. Somewhat to my surprise, he answered, stroking his rather ample front, "Well, you know, we didn't used to think so much of him but now I think he really has a place." Karl Popper said something in a little paperbook book I recommend called *The Failure of Historicism*. He said, "*trends are not laws.*" Popper directs attention towards one of the things which may help us understand diversified tastes, namely, one cannot possibly hope to look for laws. The most that one can identify are trends. Trends are contingent, laws are more or less eternal. Trends are less reliable than laws, yet understanding trends *can* lead to the laws of their development.

If we believe, as I suppose I do, that it is important to try to identify what these trends may be and to identify not only what they may be but how they got that way—how to analyze them—then we come to the next idea I want to bring to attention. That is, the question of social predictability. Popper quite rightly claims that predictability is out of the question. There are too many variables; you can build a graph of what has happened in the past but you cannot reasonably extrapolate it into the future. In any field, and this would apply stringently to the one we are interested in, there is a factor of importance today which was not nearly so evident when Popper wrote his book, namely, electronic computing. Today it is conceivable to have fantastically rapid mass records of different reactions to the same thing, or to two different things. It is possible to gather vast bodies of experience much faster, much more extensively and, therefore, to make much more interesting comparisons between such data than we have ever been able to do in the past. This is a quantum leap in our ability to understand the interaction between the purchaser and the design presented to him.

I should now reveal my biggest, tallest, and I think, handsomest hobby-horse of all. The greatest difficulty in the world of design is an extraordinary blindness to its major raw material—the consumer. Now what you like as an individual, or I, or Mr. Hoving of Tiffany's, are things that nobody can analyze or finally explain. However, average reactions, given certain situations—i.e., numbers of people in certain places, at certain income levels with certain backgrounds—do work out in averages and can be analyzed. A good designer has to know what he is working with so he can master his materials and his methods of using material. If it is metal, he has to know how to bend it. If it is wood, he has to know how to cut it and join it. If it is stone, he has to know how to amass it so as not to create a catastrophe. He has to be able to take advantage of its inherent response to gravity. And just as any designer has to know his materials and what they will do, so too must any designer in the modern world, or any man who is going to work with designers in the modern world, know his audiences, purchasers, the public or publics. That really is the prime "raw material" of the design world, much more than the physical things I have mentioned.

Now if we consider the question, what would make this, or that, segment of the public react in a certain way to a particular stimulus (to a particular model of a new car, to a particular hairdo, or to a particular line of printed fabrics—it makes no difference) then we come face-to-face with the most frequently cited resource of designers, namely, novelty. What do we really mean when say, if you come out with something "novel," you will have caught the attention of your public and they will respond to it; thus, you will have to keep on finding more and more novel things? In the long run, there are two basic attitudes about novelty. We might call the first one Heraclitian novelty, that is, the novelty Heraclites mentioned when he said you never step into the same river twice because by the time your foot is out and in again, it is a different batch of water. Let us say you put your foot in and you muddy the water. By the time you put your foot in again, you are not putting your foot in the same clear water you did at first but in the muddy water you have created yourself. That is only one of many, many ways in which seemingly similar situations alter as time goes on and as things are tested and tried.

Heraclitian novelty has in itself two aspects. One is objective, the other subjective. The person who puts his foot in the clear water is not in the same psychic state, indeed he is not even in the same physical state when he puts his foot in the second time. There are subtle adjustments, objectively and subjectively, to these situations, no matter what one does, or however carefully they are controlled; indeed the more carefully we control our observations, the more the distinction is true. Everytime we have changed the situation somewhat, it creates the second kind of novelty, which might be called the "appetitive"novelty. It is surprising that nobody, in considering visual reactions, has talked definitively about satiation. People who run perception tests (and there seem to be millions of them) report that if you look at a red light long enough, and then close your eyes, you will see a green echo. For the moment you will have burned out your red receptors and your eyes will record the opposite in order to give the overcharged receptors a rest. This is rudimentary. A parallel situation exists when we respond to other perceptual challenges—whether it be shape, depth, a sense of restfulness or of dynamism, or whatever. If we get too much of one element we desperately need a change.

Once, in a secondhand bookstore I found a little volume by a German refugee to England who described one quite normal and simple application of this physiological reaction. He wrote about taste in its literal sense—why it was that people ate the way they did. He came up with a simple enough idea. It was all very well to get excited about Oriental cooking and other exotic foods, but in the long run the traditions of Western eating, where meals were composed of relatively well established counter-balances of taste sensations, texture sensations, hot and cold sensations, all skillfully alternated, really represent the best use of the receptive apparatus of the human body in terms of enjoying the food which would nourish it. I think that he understood taste, not only in its most literal sense, but in general. All of us need to pace our taste, to alternate it, to vary it continuously in order to get the best results: whatever it is that we enjoy completely at one moment, if we keep on driving at it over and over again, is going to lose its savor.

If all these foregoing notions add up to anything, they add up to an unbelievable confusion and chaos. Out of all these attitudes, could one come up with any insight into using design intelligently, skillfully, and I suppose one might say, fruitfully? Are we not back in a situation where (until we accumulate an enormous amount of detailed research, careful comparison, discussion about testing and all those intellectual devices used by scientifically trained minds) we must rely on what I guess is called "creative intuition?" This is a risky procedure. Not risky individually, for it is the only way in which we can satisfy *ourselves*. But from the point of view of commerce, from the point of view of effective use of design in a world of mass audiences, it is very risky indeed and nearly impossible to apply in a forthright way. But design by intuition is the general way of design in our time and it has its own peculiar difficulties which are worth a look.

These difficulties of design fall, I think, under two big headings, the first of which is economic distortions. The chief economic distortions are presented by government controls within which one somehow has to work. Then there are many economic distortions dictated by commerce. For example, what incentives are given to sales people to push this or that line, regardless of its design? These incentives are not always monetary; sometimes they are vacation trips or other little goodies. There are many, many ways to encourage that crucial element, the sales personnel, to modify the reactions of their customers to the appeal of the object or services being merchandised.

Associational distortion is another huge area. Associational distortions are constantly with us, frequently on a subconscious level. "Do you know that Mrs. Smith is not wearing her mink stole anymore? It is a silver fox throw now." Or, "Do you know that the Queen of England wears a hat that looks like a bowler? Do you know that Mark Phillips wears his hair not very long?" Or, "Do you know that some of the best people in the world today believe that there never has been anything so beautiful as French furniture of the 18th Century?" Something becomes the absolute acme of what should be done, not because it is or is not the acme, but because the best people think so. Indeed this is exactly what people used to accuse myself and my colleagues of doing because the Museum of Modern

1. Associational Distortions—
Because he was a brilliant architect and remarkably progressive in the years 1895–1915, Charles Rennie Mackintosh is much admired today. His quite impractical furniture is being made anew and distributed for no good reason.

(Collection, The Museum of Modern Art, New York. Estée and Joseph Lauder Design Fund)

Mackintosh, Charles Rennie
Replica of Chair with High Back. 1902.
Ebonized wood with upholstered seat, 55³/₈ x 15³/₄ x 13¹/₄".
Manufactured by Cassina.

Art was a prestigious place, with rich trustees, and social glamour. If the Museum said this is right or that is right, or this is better or that is worse, then it was assumed we were pushing our "associational" value.

Then there are the associational distortions of suitability. What is proper in town, what is proper in a vacation house, what is proper on a train, what is proper in a car. These are all associational ideas. What has been proper, what has been the code that was established as being suitable, does not mean it can not be modified, changed, adjusted or even contradicted, but it is always based on the associational background.

So, I do not place my bets on intuition as a basis for design in the modern world. The risks are too heavy. Instead, if I were a student of commerce I would find it terribly important to try to think about the neglected raw material of design, the public. I would find it very desirable to do more record keeping about what sells and more talking with other people to discover why it sells. In other words, I would find myself at the mercy of the forever perennial governing factor, the market. The market is the mistress and always has been. The question is, how do we understand what the market is trying to say to us? That is the real key to design in our world today. It may be that we have a little time still.

Before I close, I want to take the opportunity to mention a recent meeting of an international association of industrial designers. Industrial designers, of course, have professional associations in which they exchange ideas and try to keep their ethics straight. There is such an association in this country, and in almost all Western and a good many Oriental countries as well. An international group called the International Council of Societies of Industrial Design meets every second year to elect new officers and a board, to get the feeling of their member societies as transmitted by representatives and to have a general congress with interesting events and speakers. This year they met in Tokyo and Kyoto. I was on the Board of the International Council and each of us on the Board was required to write something for publication on the occasion. I include here part of what I wrote for the Japanese audience because it considers the same proplems that I have just discussed here:

> More than a century has passed since the Western world has begun to question the human and social values of industrialization. At that time the West also began to look to Japan for confirmation of an ideal—a way of life, simple, pure and close to nature. This ideal seems embodied in certain aspects of Japanese architecture and handicrafts. Today that ideal has vanished from the West and is vanishing from Japan despite the deep piety and responsibility which many Japanese feel toward

it and many Westerners as well. By now Japan is one of the world's leaders in technology, in its development and dessimination. Already Japan vies with the West not only in progress, but also in pollution. Can two hemispheres, two cultures, two traditions, do no more for each other than to dream idealistically and to spoil realistically? We cannot afford to believe it. Today Japan can do what the West so far never could. Japan can produce technologically what is wanted by the masses of mankind by learning to listen to the voices and feelings of mankind.

Until the era of electronic calculation, no one could dare to think that the innumerable experiences of the marketplaces could be recorded, still less analyzed. So for a hundred years Western technology has mass produced in accord with the taste of a few designers more or less sophisticated, more or less cynical, more or less gifted. This was a procedure first elaborated in the academies of Louis XIV suited to a narrow community of courtiers. Now it is possible to conceive of feedback from the globe's millions of users, from huge groups of consumers in every continent, each with its own cultural and economic idiosyncrasies. The challenge lies in the process of analyzing what computers can record. To do this understandingly is the greatest task of industrial design today. The slow patient development of inherent parameters of need and desires, the true insight into today's markets, can open up a vast new world of design activity far more meaningful than anything we now know.

With these thoughts in mind, I was greatly encouraged when our Japanese hosts decided to establish "Soul and Material Things" as the theme of the meetings this autumn. They may well be pinpointing the real future of industrial design, a design not dictated by the taste and understanding of a few, nor by the short term values of commercial promotion. Instead, it is design dictated by the study and understanding of modern mankind—multifarious, unsatisfied millions whose mute reactions have at last found a voice audible to whomever first learns to harness technology for the purpose of interpreting the soul of man. So, rightly or wrongly, is the way design looks from where I sit.

The Designer and Manager Syndrome

Sir Misha Black
Senior Partner, Design Research Unit.
Professor of Industrial Design at the
Royal College of Art, London.

Two worlds have to be bridged—that of the executive in management who creates design policy and the designer who creates design. Management and designers have important responsibilities which go beyond the reaches of the business firm and the studio. Designers and management ought to share their concern for the environment and structure of society—a concern for "man as a living, mutating, organizing, and dying entity." Design is an attitude of mind and must permeate management and be more than a painting for the president's dining room or mural in the cafeteria. Lack of design understanding and awareness in our society is due to our world-wide educational system. "Visual literacy and the cultivation of aesthetic discrimination are sacrificed to undue concentration on the development of intellectual faculties."

A behavioral understanding of the social symbolism and acceptance of products are key social and marketing research tools for a designer and manager who must plan for future products.

There is a chasm between talking about merchandising and being responsible for selling, between reading about surgery and actually inserting the scalpel, between being concerned about design and turning a blank sheet of drawing paper into instructions for manufacture. I am sure that many of the recent efforts to throw some bridges across the abyss which separates those who determine business policy from those who, as designers, give it physical form have helped somewhat to a degree, but we still remain gazing at each other with the inbred suspicion of sparring partners. My intention is, therefore, to at least partially expose the potentialities and limitations of my profession of industrial design.

We both face problems which did not greatly worry our ancestors. The purpose of business, as the executive once saw it, was purely to make money; the duty of the designer was to produce beautiful objects and environments. If he lowered his sights to enable industry to increase its profits, he was almost certainly, he imagined, prostituting his art. We are wiser, in some ways, than our grandfathers even if our clearer vision has exposed new problems reaching to the horizon. The late 20th century attitude to business will be discussed later, but let me first explain what I believe should now be the concern of the designer. To be a designer, as I understand this ill-defined and much misused word, is to be conscious of, and accept, some responsibility for the physical form of our world; to be continuously aware of the shape, size, color and texture of those parts of our environment which are man-made, of the interrelationship of component parts, whether they be static or in motion, which produce a single object or a system; to be prepared to distinguish between those objects and relationships which are aesthetically acceptable and those which fall below our personal standards. "Design," as Maholy Nagy said in the late 1920s, "is an attitude of mind," to which I would add, "and the capacity for tactile and visual discernment."

But concern for the condition of our environment and the capacity (or assumed capacity) for aesthetic discrimination do not in themselves produce a designer. Concern without the capacity for implementing change is the role of the consumer as critic; to work as a designer, technical skills and experience are essential requirements. The need to master complex techniques separates designers into fields of specialization—architecture, engineering, graphic communication, the design of products for the industries based on the ancient crafts, and into many sub-divisions of these principal fields. To be a designer, in the sense that I am now using this generic term, requires not only the skills, but also the willingness to deploy them for the improvement of the environment, rather than its desecration. To this extent design is, or should be, a moral act undertaken within the constraints of the political, economic and social systems. It is a practical art, and as such different from the arts of painting, sculpture, literature and music, which can transcend the immediate present and open windows to ecstasy. The fine arts, as Modrian has said, enable those who look at, or listen to, their manifestations—to be conscious of "the union of the individual with the universe." They have been described by your American philosopher Susanne Langer as "the creation of forms symbolic of human feeling."

Design operates at a more mundane level; its concern is with man as a living, mutating, organizing and dying entity. It is here that the business executive's interests and mine, as a designer, coincide. The executive is involved in the structure of society and so am I. His desire is to improve the physical condition of millions of our fellow human beings who inhabit the world and the quality of our lives, and this equates with the ambition of all designers who have escaped despair and enervating pessimism. Business is becoming a profession, design is achieving professional authority. It is now necessary to establish a basis for understanding and co-operation so that the business executive can harness design skills to further our common purpose. I have deliberately used the word "harness," as the initiative must come from management. A designer without a client is as impotent as an actor declaiming to an empty theatre. The first step in improving design standards is for management to decide that it wishes to do so. From this board room decision much good can flow, both in financial returns to the company and for society as a whole.

I have already suggested that design is an attitude of mind. This

1. Large scale industrial catering. The staff restaurant for BP Limited. Designers: Design Research Unit. London.

must permeate management if more is to be achieved than the purchase of some paintings for the President's dining room and the commissioning of a mural for the staff cafeteria—admirable though such patronage may be. Design is inevitably an aspect of many facets of business organization. All business executives are committed to the employment of designers, the only variant being that some are conscious of what they are doing and others somnambulistic. Design exists in the company letter heading, its trade symbol, the livery of its delivery vans, its factory and administration buildings, the furniture and tableware in its staff and executive dining rooms, in the products of its factories and the containers in which they are marketed. It is impossible to be in business without making a commitment to design; even a stockbroker has a corporate identity, while those who manufacture products or provide services are dependent on buyers' reactions for their existence.

The techniques for employing and co-operating with architects and graphic designers by enlightened management are well understood. I regret only that good intention does not more commonly achieve exemplary results. This is the fault of educational systems throughout the world, a world in which visual literacy and the

cultivation of aesthetic discrimination are sacrificed to undue concentration on the development of intellectual faculties. To compen-
sate for their hereditary cataract, management should, during this
intermediate period, look for wise counsel to support or question
their aesthetic predilections. Such counsellors now exist in the
United States and in Great Britain but they should be selected with
the care and attention equal to that which characterizes management selection of chartered accountants and lawyers.

The role of the third type of designer required by production
industries is less well understood. The function of the industrial
designer is still too often envisaged as a luxury which can be
employed or discarded at the whim of management. The industrial
designer's specialization is two-pronged. Its first manifestation is
based on anthropology and ergonomics. It is concerned with the

2. Design for Control—Collaboration between marine architects, equipment engineers and industrial designers. Design Consultants: Design Research Unit. London.
(Photograph from John Maltby Ltd.)

relationship of the users or operators to artifacts or machines, be they simple domestic or office appliances, complex machine tools, agricultural equipment or control systems. It is the aspect of engineering design that determines whether a hand tool is properly shaped and balanced to ensure maximum efficiency in operation, whether the refinement of kitchen appliances is reasonably related to the capacity of housewives to understand their intricacy, whether the control system of an automobile contributes effectively to safe and enjoyable driving. The need for specialization within the product or system development team as well as the complexity of design development necessitates the devolution of specific industrial design responsibility in creating the anonymous team which is now usually the generator of new products and systems.

The education of industrial designers enables them to participate in the creative processes which are essential to product and system development. This arises partially from the structure of their curricula, which exclude the depth of study in engineering science required of the mechanical engineering undergraduate, so as to provide time for divergent thinking and developing conceptual attitudes to product and system innovation. To these is added the advantage of studying industrial design in colleges of art and design where there is the inevitable tension engendered between students of the fine arts and of the useful arts. The argument and counter argument by which the artist and the designer attempt to defend their different but related activities heighten self-criticism and set standards of personal achievement and responsibility which are crucial in developing creativity.

The second specialization of the industrial designer is his overt concern with aesthetics, with the formal qualities of objects, with shape, texture and color, and with the visual and tactile relationship of the component parts of machines and products. Separated from mechanism and structure this becomes "styling," which aims only to encourage sales irrespective of social need; considered as a refinement of the mechanism and structure of industrial products it becomes "style," as much an attribute of product design and systems of engineering as it is of literature and music. This second specialization is not the exclusive prerogative of the industrial designer.

3. The manual to guide British Railways after the designers have initiated a corporate identity. Designers: Design Research Unit. London.

(Photograph from John Maltby Ltd.)

Engineers and business executives are concerned with the total visual impact of the products for which they are responsible just as the industrial designers are—the difference is one of degree. The industrial designer has the advantage of this being one of his specializations, a task to which he is consciously devoted and for which he has been specially educated. In this way he can be usefully teamed with design engineers whose satisfaction is dependent more on the efficiency of mechanisms than on visual and tactile qualities.

Style is the signal of a civilization. Historians can date any artifact by its style, be it Egyptian, Grecian, Gothic, Renaissance, Colonial American or Art Nouveau. It is impossible for man to produce objects without reflecting the society of which he is a part and the moment in history when the product concept developed in his mind or was the creative outcome of a group sharing common attitudes and technical capacity. In this sense, everything produced by man has "Style," but this can be debased and perverted when factors other than the achievement of excellence become the dominant motivating forces. In its pure sense, style is "an aesthetic sense based on admiration for the direct attainment of a foreseen end, simply and without waste. Style in art, style in literature, style in science, style in logic, in practical execution have fundamentally the same aesthetic qualities, namely attainment and restraint . . . with style the end is attained without side issues, without raising undesirable inflammations . . . style is the ultimate morality of the mind." In his

4. The international
style—a Chase Manhattan
Bank in London. Design-
ers: Design Research Unit.
London.
  (Photograph from
John Maltby Ltd.)

Presidential address to the Mathematical Association in 1916, A. N.
Whitehead admonished his audience not to "bother about your style,
but solve your problems, justify the ways of God to man."

We must attempt to relate this purist attitude to the design of
industrial products as a whole, which seldom measure up to the
Whitehead definition of style. The world is littered with products
from the engineering industries disguising their mechanical efficiency (and sometimes inefficiency) with symbolic forms and
decorative embellishment, and which have only a marginal relationship to engineering necessity and manufacture. These products are
usually the outcome of aesthetic decisions having been made by
managers or engineers who are unskilled in making such a judgment. Subjective aesthetic design problems are constantly posed in
all design projects which allow for alternative solutions, as do all
those which are not conceived at the frontiers of knowledge and
mathematically determined. If there are five such decisions to be
made during a design development program, and if each discloses
ten alternative solutions, this can produce 100,000 design variations.
If the number of decisions to be made and the possible alternatives
increase, the number of possible design variants quickly reaches
astronomical figures. The problem is compounded by the fact that
some industrial products consciously need to serve symbolic as well
as practical needs. This is exemplified by the automobile, but is
equally apparent in the design of domestic appliances, office ma-

chinery and, to a lesser extent, in machine tools and agricultural equipment. The motoring correspondence of the British GUARD-IAN, on 10 July 1972, described the Ford Capri 3000 E: "The Ford Capri has a powerful looking bonnet, a racy tail and squashed up seating just like connoiseurs' cars that make sacrifices for speed. The Capri's sacrifices are for style. The bonnet is only two-thirds full of engine and the stubby rear makes the boot small. . . . Ford, traditionally value conscious, has produced a stylish and successful car, and the 3000 on its own terms has the panache of a car twice its price."

So long as people are not cheated by superficial design into believing that stylish elegance compensates for engineering negligence (and this is clearly not the case in the Capri, which is excellent value for money), I see no cause for puritanical objections to dressing up the ordinary with the glamour of the extraordinary. It differs only in materials and technique from making the visual best of one's personal appearance. The desire to obtain positive aesthetic pleasure and social status from the objects used by man is as old as mankind itself. A concern with the shape and decoration of man-made products is an endemic characteristic of the human race; it stretches from the decoration of neolithic pots to supporting the structure of 19th century beam engines by classical Grecian columns, from the decoration of Saracen scabbards to the form and livery of the Japanese Tokaido express train. The present need is not to disparage this aspect of style but to ensure that the formal

5. Specialized packaging—
protection for computer cards.
Designers: Design Research Unit.
London.

(Photograph from
John Maltby Ltd.)

hoskyns
systems research
testmaster   ◇›
modular software for IBM system / 360

decisions are appropriate to the object—to decide which artifacts and systems should be negative and self-effacing and which may proudly and aggressively acclaim the social and symbolic implications of their mechanical purpose. The motor car, high speed trains and television sets fall into the expressive, symbolic category; machine tools, electronic equipment, refrigerators and hospital equipment have willingly accepted a discreet anonymity in which their formal qualities are the outcome mainly of operational efficiency and economy in production. Some products which are initially expressive recede to a negative acceptance without social overtones and then later burst out again as positive social symbols. The telephone is an example of this see-saw process. In the early 1900's the telephone was both a utility and an indicator of social status; by 1914 it was hidden under dolls with crinoline skirts; by the thirties, it was accepted as a piece of domestic, commercial and industrial equipment which warranted no greater attention or commendation than its technical efficiency; by the 1970's it has again attracted symbolic, social attitudes as the availability of new types of instruments provides the opportunity for the type of personal choice and decision which can provide positive aesthetic pleasure and indicate social status. Already there are different models offered by the British Post Office. Computers are in the early stages of an aggressive-recessive cycle. They are still a source of pride to their owners but soon to be relegated to commonplace acceptance comparable with the boiler room and the air conditioning plant, now of interest only to the specialist. The automobile has moved in years from brash spaceman exuberance to a more subdued and mannerly concern with ground speed images.

In a competitive society, the need for manufacturers to be aware of the movements in public taste and its effect on sales is an essential factor in marketing strategy. The problem is compounded as attitudes towards products and systems change at accelerating speed. For a manufacturer planning a new product requiring design development, the tooling and initial production program may spread over 18–24 months and is no longer assured of public acceptance unless he can foresee what form will be acceptable in years' time by the targeted section of the market. The capacity to

6. Industrial designers in collaboration with engineers: a train for London Transport. Consultant Designer: Misha Black of Design Research Unit. London.

(Photograph from John Maltby Ltd.)

sense movements in public taste does not require clairvoyance. It requires market research (which usually operates negatively by eliminating catastrophically wrong decisions and may sometimes indicate useful lines of technical development) as well as a capacity for instinctive comprehension based on a knowledge and understanding of movements in painting and sculpture; an awareness of the attitudes of the most creative and experimental architects and designers; and of changing social values. What is happening one year in the studios and experimental workshops will inevitably later influence mass markets. The difficulty is in the discrimination between the mainstream and eccentric excursions into shallows, and ensuring that the time scale is correct. The industrial designer, by his association with the fine arts during his education, and his continued interest in their influence on public taste, is a useful ally of both management and the production team when forward decisions must be made which will influence marketing success or failure.

The need to prognosticate social acceptance, to decide whether products have moved from aggressive to recessive positions, is not only an aspect of our capitalist society. It has proved to be a requirement in the Soviet Union where public reaction to consumer products has also become selective now that the requirements of minimal existence are more easily satisfied. The U.S.S.R. has established its All Union Research Institute of Industrial Design (V.N.I.I.T.E.) to ensure that Soviet manufacturers become aware of the volatile and sometimes seemingly irrational movements in public taste on which the domestic and export sale of industrial products partially depends. The other Eastern European countries are equally conscious of the need to employ the specialized abilities of industrial designers to ensure that their engineering products reflect the style of our century and that their consumer products combine aesthetic satisfaction with efficient utility.

In this field of man/artifact relationships, the designers still operate largely subjectively—by hunches which they would find difficult to describe or justify, except by results. The considerable volume of research work on design method has been concerned with practical problem solving, and tends to gloss over aesthetic problems as not being a suitable subject for objective analysis. There is a need for more theoretical and case study of the morphology and typology of man-made objects. Until more research work is undertaken we can do little better than follow the advice of Palladio who, in the 16th century, wrote: "Although variety and things new may please everyone, yet they ought not to be done contrary to that which reason dictates."

Reason dictates to management that they should seek specialized design skills to augment their own abilities and experience. This can be done by establishing an effective design office within the company organization, by the management of consultant designers, or by a combination of both. The latter is normally the most fruitful method in large scale industry. The company design office is in close daily liaison with those other departments which utilize or affect its work: with research, product planning, production engineering, sales and publicity. The essential modifications to products which must be made at short notice, the need to watch deviations from the established house style, and the opportunity for educational work within the organization all fall naturally within the province of the inside design office. The consultant makes periodic visits to advise and criticize. He brings to one industry his experience in many ways; not only can he talk with the company designer on terms of professional equality but he can also talk to management with a freedom not normally enjoyed by employed staff.

Whatever organizational permutation may best meet the needs of a specific industrial or marketing group, one irreducible fact remains—the quality of the design program is absolutely dependent on the capacity of management to appreciate its potentiality. Management is an invisible participant in every design project. The most frequent comment in any design office is:"It's no use, *he* will never accept it—you couldn't possibly persuade him to have it." Hence a brilliant innovation is sadly filed away in favour of a mediocre solution which will offend no one, and will make only a modest contribution instead of a leap forward. Fortunately, this is not the inevitable rule. There are too many brilliant design innovations in American and British industry to justify despair. Yet, for every innovation which sees the light of manufacture and marketing, a dozen are neglected for the lack of management confidence in the capacity of their designers, and for the lack of confidence of the designers in the vision of their directors or clients.

The essential quality of design is creativity: the capacity for predicting what technology will make possible and what people will desire and need a year or two ahead. As Swift wrote:"Vision is the art of seeing things invisible," and the invisible future will be predicted by designers who have been trained to make and then test creative hypotheses. For designers to work effectively, they need the sympathetic understanding of management. I am sure that the new breed of managers will provide the oyster shell in which designers can be the essential irritant.

Around 1881, when Joseph Wharton established his now world-renowned business school, Queen Victoria wrote: "Beware of artists, they mix with all classes of society and are therefore most dangerous." Designers are concerned only with the practical arts, and the danger is thus diminished, but they are amongst those who are willing to accept a responsibility for the look and feel of our environment. And, if this involves action which disturbs those who wish only to conserve the social and environmental patterns of the past, then Queen Victoria was right and shall continue to be so.

But we are no longer isolated. The council of the Confederation of British Industries has recently accepted a report from one of its committees which reads: "A company, like a natural person, must be recognized as having functions, duties and moral obligations that go beyond the pursuit of profit and the specific requirements of legislation . . . profit alone is not the whole of the matter." The moral obligation includes a respect for our man-made environment and the elements of which it is comprised. The new generation of business executives and their natural allies the designers, can ensure that the industries under their control will improve the physical world and not degrade it. As Camillo Olivetti said more than 50 years ago: "A good business is one which *also* makes money."

I have so far talked about design as though it were a straightforward respectable activity undertaken by men and women who differ only in their technical knowledge and specialization from the business executives with whom they must collaborate. Viewed from one position this is an accurate image of the designer's persona. When design is an ordering process that changes the inventors' or development engineers' lash-up into a marketable product, or concerns itself with industrial or civic tidiness and good manners, it does not differ greatly from any other professional occupation. But this is only one facet of design which has achieved exaggerated importance in our society, because only those specifically trained to see are able to perceive the physical world and suggest how it might be improved. While the world remains sick, all who are not motivated solely by greed must combine to construct crutches so that society can continue to exist while it heals itself. Our democratic societies are, however, not monolithic. They provide opportunities for immediate manifestations of the human spirit even if development of the body politic as a whole must be intolerably slow. There are moments, even in the most staid of businesses, when a new product or service can be perceived, when a new headquarters building or factory can be planned, when an outmoded corporate image is ripe for complete reassessment. These are moments of excitement in which all those involved may share. But they are also the moments when professional competence alone is insufficient. For innovation, creative management must be linked with design creativity, when the visionary capacity of the designers subsumes their daily plain competence. Designers with the capacity for innovation are thin on the ground in all countries—but they do exist, and a place should be found for them within the structure of industry and business. Such designers may be uncomfortable companions; they may be motivated by forces which do not answer to the bridle of normal social behaviour; they may well be the odd man out in the necessarily tidy industrial hierarchy, but the odd ball often bounces higher.

Design when it is creative is not a tidy affair: it is a search for perfection in an imperfect world. But business executives will tolerate its nonconformist intolerance, will discover how to harness creativity to their practical needs—then a partnership can be established which will make their working lives more exciting (if less ordered and comfortable) and may make the products and services which they control contribute to a future which will make our present, in retrospect, appear to have been sadly inadequate.

Good Design is Good Business

Thomas J. Watson, Jr.
Chairman of the Executive Committee, IBM Corporation

Good design was "one of the major reasons for the success of the IBM Corporation over the past 18 to 19 years." Design must reflect the practical and the aesthetic in business but above all "good design must primarily serve people." Good design is a key element of corporate responsiveness to the nation and the world as well as "business survival." The three-pronged approach to good design and subsequent business success at IBM was watching Olivetti, retaining Eliot Noyes as a designer, and developing a set of corporate policies in design management.

After entering a business which, fortunately for me, was headed by my father, I eventually came to see design become one of the major reasons for the success of the IBM Company over the past eighteen or nineteen years. We had relatively good design in the days before I ever got there, but one night in the early 1950's, as I was wandering along Fifth Avenue, I found myself attracted to typewriters sitting in front of a shop window. They were on stands with rolls of paper in them for anybody's use. They were in different colors and very attractively designed. (In those days you could have an IBM typewriter in any color as long as it was black, as Henry Ford said about his "Tin Lizzie.") I went into the shop and also found attractive, modern furniture in striking colors with a kind of collectiveness. The name plate over the door was Olivetti.

Subsequently, I went to Italy and met Mr. Adriano Olivetti, one of the great industrial leaders of Italy. He had a completely organized design program that included company buildings for employee housing—which was popular in Italy at that time—as well as Olivetti offices, products, colors, brochures and advertisements.

Shortly after this, in 1955, a close IBM friend of mine, manager of our IBM business in Holland, sent me a very thick letter in which he said: "Tom, we're going into the electronic era and I think IBM designs and architecture are really lousy. I've collected a lot of Olivetti brochures and pictures of their buildings, as well as brochures and pictures of IBM. Put them all out on the floor and have a look down each column and see if you don't think we ought to do something." The Olivetti material fitted together like a beautiful picture puzzle. At that time we didn't have a design theme or any consistent color program. All we had were some very efficient machines, not too well packaged, and some competence in the new field of computers. In fact, we were building our first family of computers—the 700 series. They worked on vacuum tubes which seemed from the inside design to be the very epitome of modern technology. We thought it was time for the outside to match the inside. That was a design problem. We took all of the top-level people in the IBM Company to a hotel in the Pocono Mountains where we considered IBM design in contrast with that of Olivetti and a number of other companies. We wanted to improve IBM design, not only in architecture and typography, but color, interiors—the whole spectrum.

The only person in my experience who knew anything about design was Eliot Noyes. During the war, I had become interested in gliders. Eliot Noyes was head of the glider program in the Air Force and we had flown gliders a few times together. After the war he became a prominent industrial designer, so I had asked him to join us in the Pocono Mountains. At the end of three days, he convinced us to do an about face in our design trends. From that day to this, Eliot Noyes has given fifty per cent of his time to the IBM Company, never as an employee, always as an independent consultant. It has been a wonderful relationship.

With Eliot's arrival, we organized our design plans. We had just three factories in those days. In every IBM factory and laboratory today, there is a design section free to change the exteriors of our machines, if it does not hinder their function, in order to make them fit a cohesive and attractive design. It is done autonomously at each lab and plant, but always under the general supervision of Eliot Noyes. Eliot also travels, lectures and advises people to come to the various centers of design here and abroad and to keep their ideas modern and fresh. Furthermore, as we design these machines, we are aware of our position at the cutting edge of technology in electronics, a technology which can often be physically beautiful. The actual mechanisms themselves make lovely pictures, so we finally put in safety glass and let the customer or observer look into the machine mechanism itself, rather than try to hide it under a cover. At the same time we began to work on good office and show room interiors. We have used probably hundreds of interior decorators over the past twenty years.

After Eliot Noyes, we took on Paul Rand to work on design and Charles Eames to do films, exhibitions and museum activity. Charlie has designed the exhibit called "Computer Perspective," which was on display in our Manhattan office building. He knows how to explain computers to the public. Charlie can put what a computer does into a little cartoon-like film and in the course of twelve minutes have everybody in the room understanding the main computer functions, what the world is looking for from computers, and how they work.

In the course of fifteen years, from 1956 to 1971, we built about 150 plants, laboratories and office buildings. Whenever we had one to build, we would get the names of three good architects from whom our own design people would pick as appropriate for our needs. The names that we picked from were familiar names: Mies Van Der Rohe; Breuer; Eero Saarinen; the late Egon Eiermann of Germany; Jacques Schader of Switzerland; Marco Zanuso of Italy (who is an excellent interior decorator as well as a designer); Jorgen Bo of Denmark; Sten Samuelson of Sweden; Shoji Hayashi of Japan, and the late Henrique Mindlin of Brazil.

What has been pleasing about our design program is public response. People would return from abroad and say, "What great

buildings you have around the world!" One day Mrs. Lyndon
Johnson, who had seen our building in Hawaii, telephoned to ask
me, "Will you come down to the White House and tell us how you
achieved your design program in the IBM Company? I'm trying
to beautify and change the way Washington looks." I said, "Mrs.
Johnson, I'd love to come down, but the brains behind this project is
Eliot Noyes. Can I bring him?" I did, and we spent a Saturday at the
White House. Those delightful daughters of hers came into lunch
with their hair up in curlers, underscoring the informality of the
occasion. We had an absolutely enjoyable time, which I credit to our
design program.

When I think of Noyes and electronics, I think of a story about a
fellow who had a genius for a son. The boy was interested in
electronics, so at Christmas time, the father went into the best
electronics supply store he could find and told the story of his
eleven-year-old to the salesman. The salesman said, "I think I have
what you want. We have a color TV set here which is all in pieces,
which the boy can assemble." The father said, "No, he assembled
one of those at age seven." Then the salesman said, "How about a
single side-band radio station?" The father said, "No, he had his
license for that at nine." The father wandered through the store and
came to three huge trunks, full of transistors and cores and all
varieties of monolithic circuitry components. A fourth trunk was
filled with wiring diagrams. He said, "This may be what I'm looking
for." The salesman agreed, saying, "This will teach the kid about
life. No matter how he puts it together, it won't work." Eliot Noyes
taught us a different lesson. He knew how to put together our design
program and we have the proof that it does work.

What is the definition of a good design in the IBM Company? We
feel that good design must primarily serve people, and not the other
way around. It must take into account human beings, whether they
be our employees or our customers who use our products. Our
machines should be nothing more than tools for extending the
powers of the human beings who use them. As a consequence, our
design, our colors, our building interiors are intended to comple-

ment human activity, rather than dominate it. Naturally, we are interested in the cost per square foot of a plant we intend to build, but we are equally interested in good design. We try to balance the two considerations. We also know that you have to pay a premium for good design, but that premium is paid back as many different benefits to the corporation in its activities.

A good architect wants to experiment, to pioneer. There is always a dialogue—even a conflict—between a good, strong-minded architect and a good, purposeful company. We have had some lively conflicts with some of those better known architects. For example, we have a plant in Boca Raton, Florida, done by Marcel Breuer. It was the third building he did for us. He has a perfectly delightful personality. He appears to be disarmingly soft, but he is like a piece of steel inside. Part of his intended design included a large lake in front of this building with an island on it and some mobiles on the island. I thought it was a great idea but it was going to add as much as $600,000 to the cost of the plant, which was already about $40 million. About that time the country also went into that economic recession of 1969 and 1970. So I had to eliminate the island, a decision which almost lost me my friendship with Marcel Breuer, which I prized highly. I made a promise to him that at some point in the future the island would be installed along with the mobiles. If any of you visit that plant in Boca Raton, feel free to visualize an island in the middle of that lake. It will be there one day.

There has to be a certain amount of conflict in such matters, and one has to cajole, persuade and even insist that the architect move only a reasonable distance beyond the last best thing he has seen or done. I have just been in Rome and have seen plenty of historical reasons for not letting architects get too far ahead of you. You may remember that the Renaissance Pope, Julius II, had his problems with Michelangelo. (It can be argued that Michelangelo had his problems with the Pope, too.) For Julius II, he designed a sarcophagus, larger than Saint Peter's basilica in which it was supposed to rest, and so a new basilica had to be planned. The time schedule got completely out of hand, and Michelangelo lost some of the support of the Pope. In the IBM Company, we would call that a loss of both cost and account control.

If a corporation decides to be a design leader, it must have a good advisor, which is why you are hearing the name, "Noyes" frequently throughout this lecture. Without a good advisor, the design program may be garish, or what designers call "kitsch," as a result of trying to go a bit too far. Experimental design, carried beyond disciplined control, often becomes nonfunctional, wasteful, and expensive. Good design has to meet functional requirements. It has to serve as good background and be subordinated to the human and machine activities it supports. It ought to create a pleasant atmosphere, whether it is a building, a computer, a piece of furniture or an interior. In all cases, the design plan is there only because people are there.

We built, for example, a very modern monolithic circuitry plant in Endicott, New York. In this plant, production was on a highly automated basis with very few people and highly mechanized production lines. The factory interior looked almost like a modern drawing room. While building it, we decided the cafeteria of our oldest plant, also in Endicott, was outdated and that we would put a cafeteria in the lower floor of the new plant. The beautiful theme of monolithic circuitry production was projected into the new cafeteria. There were three tiers for the tables and unusual colors on the walls. The place looked unbelievably clean, but in our other Endicott installations there are many oily, greasy screw-machine type operations. When that cafeteria was opened, many of our employees from the older plant operations refused to come in and eat because they were afraid, and embarrassed, that they would soil it. It was over-designed and we had to "unplush" it. It caused an uproar. Many of our oldest IBM employees are in Endicott and most of them own IBM stock. When they see money being wasted, many write immediately to me saying, "Somebody has lost his mind here in Endicott. They're taking the shrubbery and plants out of this lovely cafeteria and tearing down the walls." But we finally converted it to a place which was designed appropriately and was comfortable for people. Only then did it become popular. This is a good example of how over-design wastes money.

Design in industry usually encompasses a mixture of the practical and the aesthetic. Even the way an organization is designed can determine whether it is ugly or beautiful. If it is well designed, it can respond to the future. It can change its form and it remains competitive. But if it is rigidly designed and inflexible, an industry can go out of business within a few decades. Certainly good design flavors the relationship of a corporation with its many publics—its employees, its stockholders, its customers, its social critics and the multitudes of business watchers.

My father headed IBM from 1914 till his death in 1956. Because I worked a number of years for him, I can tell you some of his goals for the IBM Company which relate to our design program. Business growth alone was far from his top objective. What he wanted was to win a place for IBM in the estimation of people, and he realized that we had to earn it not only by what we did, but also how we looked. So we wore white shirts and dark suits and even stiff collars. Father felt that non-conservative dress might confuse a sales prospect, whose mind might stray, even enviously, away from the product that the salesman was trying to sell, to the cut of the salesman's coat or the design of the shoes. I did manage to break away from the stiff collar. We now wear more comfortable collars. People still smile at our dark suits and white shirts. But we in IBM smile along with them. The stockholding public also smiles happily at our growth and success.

Long before we had enough money to launch a design program, we tried to look more successful than we really were. We had an unusually fine showroom on Fifth Avenue by 1926. I remember standing on the balcony to watch the parade for Lindbergh after his flight to Paris in 1927. I am sure if we used the same percentage of our earnings today that we used on that Fifth Avenue showroom, we would own about two blocks on Fifth Avenue. And yet it caused people to look up and say, "What do those letters 'IBM' mean?" The windows helped get our products known and this in turn helped to sell them.

I would like to illustrate some of the points that I have tried to cover about the evolution of design in our company, using the following illustrations:

1. This is the parachute designed by Leonardo da Vinci—a design well before its time. Leonardo wrote, "If a man have a tent made of linen of which the apertures have all been stopped up, and it be twelve braccia across and twelve in depth, he will be able to throw himself down from any height without suffering any injury." Except for its shape, Leonardo's design resembled the modern design.

(Model from a sketch by Leonardo da Vinci. Collection of IBM Corporation)

2. Our old Endicott plant. It was begun in 1902 and sort of evolved into this look. And it sort of looks as if it evolved.
(Collection of IBM Corporation)

3. This assembly building in Rochester, Minnesota, was designed by Eero Saarinen. This one actually cost less than contemporary factory design that we were following elsewhere with people less creative than Eero. It's pretty rugged climate out there. The plant is now fifteen years old and it's as good as the day it was built.
(Collection of IBM Corporation)

4. Our first product development laboratory didn't lure many engineers to come and work for us. When I returned from the Air Force in 1946, we had only four degree-holding engineers in the whole corporation. I think improved laboratories helped attract many others, including Dr. Isaki, who just shared a Nobel prize.

(Collection of IBM Corporation)

5. This is our Research Laboratory in Yorktown Heights, New York, also by Saarinen. Dr. Isaki works there. It's built of fieldstone and the front is dark glass. Before building a new structure, it is smart to make some mock-ups. The engineers said, "We don't want outside offices. We want concentrated production and scientific talk between offices. So we don't want any windows to distract us." So there is only a large corridor inside that dark glass and every time one moves along it, he gets a fine feeling from looking out at the lovely countryside through that glass. We mocked it up and had a couple of engineers work without windows and it didn't bother them. That lab is fifteen years old now and it still works very well.

(Collection of IBM Corporation)

6. The lab that excites us the most is at La Gaude, just above Cannes on the Riviera. For anyone who would like to work in telecommunications research, and qualifies, I think we could offer a very exciting work environment here at La Gaude. It is only twelve miles from Monte Carlo and seven miles from Cannes. At least that appeals to a lot of the French IBM engineers at La Gaude. (Collection of IBM Corporation)

7. This is the last building that Mies van Der Rohe did. It houses all of our offices in the Chicago area.
    (Collection of IBM Corporation)

10. The building block of the early IBM computer is kind of hard to believe. It was a machine that sorted cards for the 1910 Census. We now sort cards electronically, about a million times as fast as that machine. But without that machine, even with its crude design, there would never have been an IBM Company.
(Collection of
IBM Corporation)

8. (opposite, top) This is the electric typewriter after Eliot Noyes redesigned it. It may not look very modern now, but in 1952 it looked awfully good and it got our typewriter business out of the red where it had been for twenty years. Only a product with good design, plus good function, can be called excellent. This IBM electric has the same insides as the earlier one, but the Norm Bel Geddes firm, with Eliot Noyes, put a new frame, or cover, around it, and it really helped.
(Collection of IBM Corporation)

9. (opposite, bottom) Our newest typewriter—the IBM Selectric.
(Collection of IBM Corporation)

11. This is a modern computer room representing a span of fifty years from the previous illustration. I do not want you to think this machine evolution happened overnight. But, we are rather proud of it as a total machine concept. This example is a 360 Model 85, which is pretty close to the top of our previous line. (Collection of IBM Corporation)

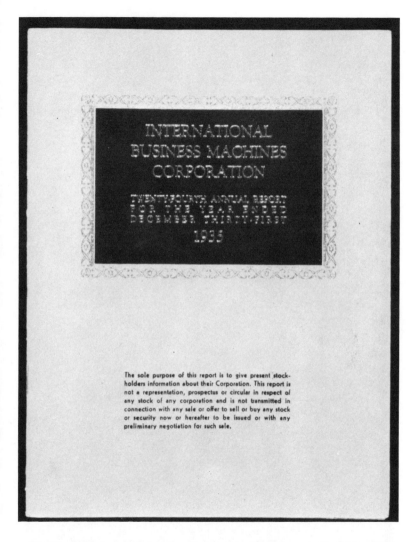

The sole purpose of this report is to give present stock-holders information about their Corporation. This report is not a representation, prospectus or circular in respect of any stock of any corporation and is not transmitted in connection with any sale or offer to sell or buy any stock or security now or hereafter to be issued or with any preliminary negotiation for such sale.

12. We did not change just the design of our machines and buildings. This is an old IBM annual report which I guess somebody tried to make look like a Venetian manuscript.

(Collection of IBM Corporation)

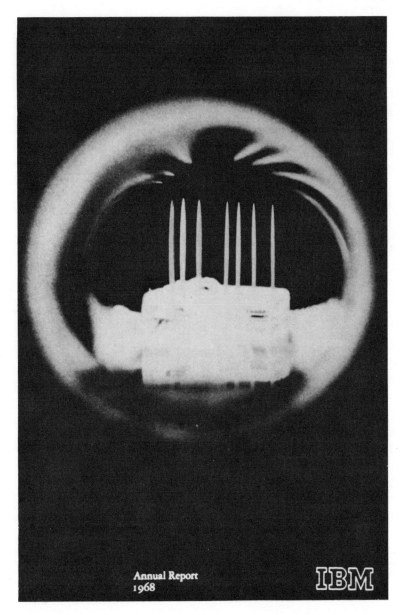

Annual Report
1968

IBM

14. The old IBM logotype, which didn't have much excitement to it.
(Collection of IBM Corporation)

15. This is our new IBM Logotype, designed in 1956 by Paul Rand.
(Collection of IBM Corporation)

---

13. (opposite) This is a more recent IBM Annual Report. We frequently
display photographs of miniaturized electronic circuitry on the front of
those reports and they make quite artistic and colorful covers.
(Collection of IBM Corporation)

16. I use the IBM Netherlands laboratory to illustrate that each of our modern buildings are quite different from one another. We think that if you are building a number of different things it is a good idea to use a number of different architects, in this case, SamenwerKende Architectenbureaux. Nobody has a monopoly on creativity. (Collection of IBM Corporation)

17. Our German Headquarters in Sindelfingen, Germany, is just outside Stuttgart. Our German company started after the war with a lot of secondhand IBM Army machines, left there by both U.S. and German armies. They both used our machines, you see. Last year IBM Germany went over the billion dollar mark in gross earnings. It uses good design (by Professor Egon Eiermann, Architect) and clearly that helps. (Collection of IBM Corporation)

18. On the other side of the IBM Nordic Education Center in Stockholm is a great, open fjord, part of the entrance into Stockholm Harbor. It is a fine example of finding a dramatic location, and putting an exciting building on it to enhance the environment in which customers come to learn about computers. The design helps to ensure a stimulating experience for guests. Every window looks out across the great body of water.
(Collection of IBM Corporation)

(Collection of IBM Corporation)

I think it is relatively easy to measure the impact of design on a product. As I said, we could not get our typewriter division out of the red until we had it packaged in a way that made typists want to use it. Their hands have to provide only an ounce of pressure instead of the seven ounces that a manual typewriter requires. But it should also look attractive on her desk.

Design is good business in countless other ways. How much business did a good looking exhibit attract to the IBM Company? To what extent do good looking facilities invite people to apply for work at IBM? These are intangible things that we believe are genuine dividends of a good design program.

Everybody wants to be a part of, or do business with, a winning organization. If the organization looks the part with good design, people begin to think the company is going somewhere in the world. History alone is the ultimate judge of what good design is. We can only act on our tastes and instincts, temper them with business consideration and hope for good results. The Egyptian pyramids survived because of sound engineering. But I think one reason we go to see them is their design—very simple, very attractive, very pleasing to the eye.

We remember the merchant cities of Venice and Florence, not because of their sophistication and wealth, but because of what resulted from that sophistication and wealth in the arts and in design—in painting, architecture, pottery, sculpture. Nobody can consciously design for posterity. But it behooves anyone in business today to pay full attention to the value of design, as long as he means for that business to go on serving the nation and the world in all of the relevant ways that spell business survival.

On several occasions I have summarized my belief in design as a strong business success force: "In the IBM Company, we do not think that good design can make a poor product good, whether the product be a machine or a building or a promotional brochure or a business man. But we are convinced that good design can materially help make a good product reach its full potential. In short, we think that good design is good business."

The Environment for Creating Good Design

Van Day Truex
Design Consultant

The key element for appreciating and understanding good design is "training the eye." Management has been too preoccupied with the problems of "will it sell?" and how products are made. The important question should be how is the product to look? Mr. Van Day Truex reflects on how Paris inspired him to relate design quality and society. The utmost in beauty and quality design continued from the Renaissance to the Industrial Age or early 19th Century—". . . that remarkable combination of form following function." The reasons for bad design are two-fold: an insufficient number of trained designers able to share the responsibility for design quality, and "very questionable managers who refuse to accept responsibility for design. . . ." The history of design shows us that "nearly every design has already been done." Design ideas and reeducating the eye can come from understanding the past. "The only really different ideas in our contemporary world are appearing because of a new material discovery or a new means of evolving a shape."

When I wish to convey the clearest distilled statement of how one trains the eye to appreciate better design, better quality, it may serve a purpose if I rapidly go through my own training or my own experience that landed me in this sphere of influence. So many aspects or angles of my experience and education over the years have been related to the cultural and design activity around me.

I had absolutely nothing in my favor but a weak desire to know more about the world of art. I was born in a high wind in Kansas. My relation to the merchandising world began because my father was one of the first members of the Golden Rule stores which became the J. C. Penney Company, created in the small towns in the far west. And my father was one of the early managers who was sent to a little town near the Jackson Hole Country at the base of the Great Tetons. My first contact with the merchandising world was I think at the age of six. Mr. J. C. Penney himself came to conduct the yearly sale and I was paid 10¢ to deliver the bills of sale around the little town. It was a nice spring day as I set off with my bills and I got lost in the sage brush that surrounded the town. The first flowers of the spring, the dogtooth violets, were coming up. As I put down my bills of sale to gather the violets, a shadow loomed over me. When I looked up, there was Mr. Penney, Mr. J. C. Penney himself. He then gave me a lecture on the importance of work—when you work, you work, when you play, you play. He helped me gather my bills together and I went on to distribute them. About half a century later, Mr. Penney, who was still vigorous at 90 years, walked unannounced into my exhibition of drawings which was being held at a New York gallery. After arriving, he made a very careful inspection of the drawings and as he left he said (I had forgotten about our encounter over half a century ago), "I was the one who was wrong, not you. You weren't cut out to be a merchant!"

From the far west and my early childhood there, I went to New York to the Parsons School of Design. It had the only catalog I could understand in my yearning for art. I had been called "artistic" because I painted bouquets of dried milk weed pods in silver and gold and had given them to my mother. I used to make tracings from photoplay magazines of the movie stars of the moment. It was my beginning in the world of art? The next thing was the Parsons

School where I stayed for thirty years from the time I entered as a student until I left as the head of the institution. So I had a long period of training before the age of 50 when I decided to leave.

My personal involvement with "enlightened" management began almost immediately after leaving the Parsons School. Mr. Gilbert Chapman, who was then the head of the Yale & Towne Manufacturing Company asked me if I would be interested in doing something about architectural hardware. At first I was dumbfounded because I did not know *how anything worked?* I was completely ignorant about mechanics. So I thanked him for asking me but said, of course, it was impossible. He listened politely and said, "We will think about this further." He said that at Valley Forge the industrial design division was staffed with over forty industrial designers, all of whom were efficient and excellent technicians, but maybe lacked a visual approach. His concern was how architectural hardware *should look*. He turned to me and said, "Think what hardware was in the past. Now look at it. It has lost all sense of adornment; cannot we have a marriage of *function and embellishment?* I want something done." Intrigued by the assignment, I asked if I could call everything a "whoseit." He said of course, "Call it anything you wish." So with that encouragement, I embarked upon this program backed by the president of Yale & Towne.

Two years later we staged a show at the Wildenstein Gallery in New York. It was a fascinating program because it crossed the borders between industrial design and the fine arts which was a new approach that was not always given immediate approval. When I had gone to Mr. Lipchitz to ask the great sculptor if he would design a door knob, he said, "No, I am an artist in the fine art sense, I couldn't do a thing like that." I didn't give up. When we had lunch again, I said, "Mr. Lipchitz, if you will not design a knob or a hand lever, would you do a symbol of welcome that might hang on a door?" He was excited and agreed to design a symbol of welcome. The final exhibition shown at Wildenstein was composed of works done by some of the finest artists alive. It included works by architects such as Philip Johnson and designs by Leger, Miro, Spadini, etc. which were executed in ceramics. Two beautiful knobs were also submitted by Noguchi. There was a mixture of architects,

1. 18th century Chelsea
Botanical Plates repro-
duced at Mr. Truex's re-
quest for Tiffany & Co.
(Courtesy of
Tiffany & Co.)

artists and industrial designers showing about 200 different items in
the show ranging from the predominantly functional, to works of art
as such, and to the frankly decorative.

In the reshuffling which occurred during the next two or three
years at Yale & Towne, Mr. Chapman resigned and the concern of
the top management shifted from excellence of design to the more
practical question of "will it sell?" This shift represents well the
opposing forces which cause often a tension at the very top of
management. Even though in this case production won a higher
priority than design, our program had been the most successful of its
kind—too bad it had so little future. It was at this time that Mr.
Walter Hoving asked me to come and see him at Tiffany's. He asked
me if I would be interested in the china and glass and silver
departments. Once again, I said I did not know anything about
china, glass and silver. He said, "Don't worry, because the buyers
know all about quality of material, but they had lost all sense of
design and imagination. If you will accept the challenge, I want you
to approve of everything that is done in design. You are not to think
what it costs, that is not your business. Your only concern is quality
of design!" From the beginning, it was a challenge.

For one of my first projects in another field, Mrs. Vincent Astor wanted a large carpet done for the new library that they were building on the upper Hudson estate. She asked if I would design a carpet for the big library which all of her animals used, the cats, dogs. It must be something that would conceal the spots of the livestock that would invade the room. It was a fascinating design challenge and I worked with Edward Fields of Custom Carpets. Together we went into the workrooms which were filled with the projecting machines they use for driving the wool. As I studied different patterns, I thought about animal skin markings, which created a spotted kind of pattern resembling a Jackson Pollack painting. The entire design was done in the natural tones of animals, the beiges, off-whites, yellows, browns and blacks. Apparently, it was successful, because over a period of four years I received a considerable sum in royalties from that one design.

Now the last challenge is to design in one *certain* medium. I was requested by Baccarat to design their internationally known crystal.

2. An array of silver and vermeil boxes designed and adapted by Mr. Truex for Tiffany. Design inspiration for the textures came from such diverse sources as oranges, rope and rice.
(Courtesy of Tiffany & Co.)

It is really the first time that I have had to concentrate on using one certain material in designing. There are two distinctly different approaches that can be taken. The designer who approaches the challenge from the material aspect whether he is working in wood, metal or whatever, must first question the relationship of the material to the product. My own concern, however, is not how a thing is made, but rather how it looks. It was the same in art history. I was never interested in dates or the specific data of any object unless it just caught my eye for some reason. In my work as a designer, I have always remained aloof from the workrooms, from concerning myself too much with how a thing is done. I have never been influenced, aided or limited by technical knowledge, which perhaps is why I could ask for certain results nobody had previously captured. I have also learned, in working and designing with one material, that if your knowledge of the technical possibilities of crystal becomes too pronounced, you suddenly may become so excited about a certain result technically, perhaps in the construction of an object, or the way you attach a handle or a supporting stem, that you forget to question *what it looks like!* Two of the most successful designs in the present Baccarat collection prove this point. They are, from the point of view of my design judgment, at about the lowest aesthetic level, but they are technically brilliant and are also selling well. One is a water goblet with its stem pulled off center. It looks unbalanced, as if it were about to tip over. Another example of poor design is a glass which has an attaching stem that is so thin it looks as if it couldn't support the weight without breaking. The fact that it does not is an extraordinary achievement technically. The design, however, is inferior.

To complete my education in design, I went to Paris at the end of my second year of Parsons on a scholarship. Two other applicants ahead of me turned it down so I was the second runner-up. I went in 1925 and stayed fifteen years. I cannot possibly emphasize what it meant growing up in Paris, the cultural and aesthetic center of Europe. For fifteen years I was in incubation in a city that was the center of style and fashion and where there was a very interesting marriage between the private taste of leaders of fashion and the designers. There was a very potent ambience in Paris which brought men and women in society face-to-face with the designer. As we emerged from the Dark Ages into the Renaissance, there was almost nothing really ugly in Europe until the beginning of the 19th Century. All the simple instruments, the metal works and artifacts used in farm activity were beautifully functional objects constructed by peasant artisans. As you rise from the simple implements used for tilling the soil to the sophisticated objects of the ruling cultural aristocracy, you find that remarkable combination of form following function. Until the early 19th Century there was the purely functional utilitarian aspect of design, which is why many collectors have studied those beautiful early implements. Then came the influence of the ruling aristocracies which preserved and enlarged a sense of the functional. The things are still beautiful even though our life has changed. The Versailles of Louis XIV was *functional*—in its splendor it impressed and manifested the pomp and glory of the *Sun King*. The baroque interior of Weis, that small late-18th Century church in the fields outside Munich (stern exterior—interior a galaxy of rampant imagination, flights of fantasy in a sunrise scheme), was *functional*—it lifted the spirits from the harsh aspects of life in the rural 18th Century community.

Until the beginning of the 19th Century industrial era, this mainly functional aspect remained intact—whether at the peasant level or the aristocratic. With this 19th Century industrial change, however, much of the quality control ceases and we embark on an undisciplined indiscriminate copying of the past. Questionable "originality" evolves out of the urge to produce to sell—good–bad–indifferent takes over. Functional progress continued, of course, in a technical sense, and at an increasing pace—inventions revolutionizing production etc., etc. But as technical progress gained, the aesthetic waned. The artisan disappeared—mass production took over!

The happy combination of hand production—on peasant and aristocratic levels—ceased. This civilized union disappeared in the morass of industrial development and technical advance.

Now *maybe* it is not too late to halt the lack of design quality that characterizes too much of mass production. But it can only be achieved by increased cooperation between enlightened management and equally enlightened designers. Without both the future is

3. Borrowing from nature, Mr. Truex designed these sterling silver objects for Tiffany. Clockwise, they are: a cabbage tureen, a box inspired by a seed pod, a pine cone box, and the seed of a tropical fruit magnified 200 times.
(Courtesy of Tiffany & Co.)

grim. And I stress first the responsibility of the designer. If he produces good design, then we have only good design on the market. If he compromises (or knows no better!) management is less guilty. But, to *stress again*, enlightened quality-seeking management becomes more and more important in keeping designers properly challenged.

I have been lucky in my search for the improved eye: Parsons training, the years in Paris between the wars and my association in private life with individuals of design judgment, top designers and artists in the professional world—years of observation and appreciation, aloof from the pressures of production and fashion, both of which are *constructive* if disciplined.

It was in this new period of industrial growth that Parson founded his school. He was a staunch Bostonian with a stiff manner of speaking and a stiff mind. He realized that this overwhelming industrial development had proved an enormous impact on the world of design. For various reasons, ranging from greed to stupidity, there was, and has been, much too much production and much too much designing. Even worse, there is too much bad designing because there are not enough trained designers who are humble enough to acquire the responsibility for quality of design, and certainly, because there exist very questionable managers who refuse to accept responsibility for design. As a designer, I am prone to blame designers. Too many of us consider that we have a good design sense without developing a continuous sense of observation to keep one's eye better informed and better instructed. Those who are training for management can help themselves by learning more about what looks right. Too many designers try to be *original,* to do something *different* at the price of achieving a poor quality of design. The more one looks at the history of design, it becomes readily apparent that nearly every design has already been done. The only really different ideas in our contemporary world are appearing because of a new material discovery or a new means of evolving a shape. New technological possibilities can open up an area and give a sense of freedom to design. As far as shapes and inspiration, however, it has all been latently touched upon or expressed in concrete form in the past.

In the little Etruscan museum in Volterra there is a bronze figure that is pure Giacometti. This is not any adverse criticism of Giacometti but, instead illustrates the point that any sensitive artist or sensitive designer uses the past. Giacometti was probably not aware of that elongated Etruscan figure when he visualized the shape in his mind. He was an artist and it became a part of his own vocabulary. Top artists understand the past. No one more than Picasso values and has fed himself more on the past, from the paintings and decorations in Heracleum and Pompeii, to the art of Africa. If I ever find myself devoid of ideas, I waste no time in getting myself to the Museum of Natural History or I walk into maybe the armor section of the Metropolitan Museum, or I look at the musical instruments. In each section, you reeducate or refurnish the eye.

I hope nothing I have said about design diminishes the importance of function. Quoting again Mr. Sullivan's dictates that form follows function, I would add Mies Van der Rohe's "Less is more." It is the responsibility first of the designer. At the same time, no good cook remains a good cook if the mistress or the master is not there to encourage and appreciate the quality of the cooking. The same is true of any designer. The designer is at his best when he has understanding, cooperation and direction from the top of the organization.

Design for America's Third Century

Nancy Hanks
Chairman of the National Endowment for the Arts

The initiative and strength for good design in America's third century will come from the private sector. The forces giving rise to a new emphasis on design in America are: corporate design improvement programs; government programs for a Federal architectural, interior and graphics review; and the increasing sensitivity of society to the quality of life and urban environment. The process of design involves the designer, user and client. The key to good design is in the hands of our business leaders—"they have the decision-making power to initiate good design and designers." Too often, design is left to the lowest levels of corporate management. And, often, design is viewed as 'decoration,' 'frill,' or 'superficiality.' "The majority of our products, buildings and graphics have unfortunately happened by accident." Part of the blame can be laid to the unenviable fact that the United States is the only nation of the world where there is no government program for encouraging improvement in industrial design. Federal and state monies, however, are increasing significantly for programs which will help improve design in our cities, states and government. "What we build and what we produce and where we live should reflect the highest standards toward which each of us aspires. And the sum total will be our design for America's third century."

Anyone attempting to predict the state of design in America's third century might be considered divinely informed or hopelessly foolish. Nonetheless, I wish to prophesy that design—good or bad—is going to be a critical factor in America's third century. And, further, I predict that it will be good. My confidence as to the importance and the quality of design is not based, by any means, on Federal action by itself. Rather, private initiative and private action—and knowledgeable citizen concern—are the key factors.

At the moment, however, the United States could be regarded as a "developing country" where design matters are concerned. Fortunately, many signs point to a significant change of thinking about design in the country. In the private area, we see some major corporations, for example, moving toward design improvement. The very establishment of the Tiffany Lectures at the Wharton School is evidence of a growing awareness. In the government, the fact that the National Endowment for the Arts has a program in Architecture and Environmental Arts is significant—especially when one notes that in the early years of the Endowment, architecture was paid scant attention. Today, it is one of the most important areas of emphasis of our advisory body, the National Council on the Arts.

There are many reasons for the nation's change in thinking as to the importance of design in all of its aspects. In part it is due to a growth in our population; in part it is a growing concern with what in parlance is called the "quality of life." In part it is because the increased complications of an urban society require better communications.

And, in part, it is due to the fact that we are a people who have been painfully matured by the fast pace of events during our lifetimes. We have seen problems assume frightening urgency within a few years. We have seen solutions tried —some successful, some failures. In this accelerated-time-and-event world, it is often said that we do not have the luxury, as our forefathers did, of wondering about the future consequences of our actions. From a purely selfish point of view, if for no other, that is sheer nonsense. For we ought to be more concerned about the consequences of our

1. Design is not a luxury, but a necessity.
Project: St. Francis Square Housing Project. San Francisco, California.
This housing project reflects the important role design played in achieving a pleasant, open and safe environment for a large number of families.
    (Courtesy: National Endowment for the Arts)

actions because we are, in a real sense, participants and witnesses, at the same time, to our future.

We no longer can—or do—delude ourselves with shiny visions of the future that will one day miraculously appear intact, somehow divorced from our everyday actions. We know that we are linked inextricably to that future; that what we do today will be a part of the future that we, not some following generation, must live with—and live in.

And what is the process by which we build our cities, build our environment, build the world in which we live? In the future, as in the past and as at present, the process involves the designers, the users, and the clients.

Future designers will continue to amaze us with new ideas and arrangements of form, just when we are sure all fields have been explored. And occasionally, they will fall on their faces—for that, too is the nature of exploration.

Users will be with us too, and will continue to have real and imagined consumer needs to be met.

If designers and users are constants in an equation, then the *clients* become *the future variable* which can create the situation in which "good design" can flourish. It is the clients who are in the unique position to insist on—and demand—design excellence and quality.

Today's business students (tomorrow's business leaders) will play a very important role in the design process. They will be clients as well as users; they will be decision makers.

Decision makers in both the public and private sectors in this country have for too long shirked their responsibilities as design clients. To some, the word "design" connotes decoration, superficiality, or even femininity. Others simply do not think about it at all. Too often, the responsibility for design decisions slides downward from one level of management to another until the decision is made by the least equipped member of the staff or it is made by default.

A recent Canadian publication stated that "everything that doesn't happen by accident happens by design." Because of management's great lack of concern for design, the majority of our products, buildings, and graphics have unfortunately "happened by accident." For that reason, companies make products that could be improved (or are not needed); people wander around confused by poorly designed graphics; newsletters and publications—totally ineffective as communication devices—are printed by the ton (using paper and trees for no purpose); and buildings are constructed, then torn down because they ill fit changing functional needs or did not take into account social requirements. I could go on and on. The list of inefficiencies alone, not to mention the lack of aesthetic satisfaction represented by our long neglect of design responsibilities, would fill a very thick catalogue.

However, from the vantage point of the National Endowment for the Arts, we have seen a very definite shift in attitude toward design—by client and user alike. There is an obvious awareness that is growing and being shared by many people as to the importance of design.

Before speaking specifically about our experience in architecture and the environmental arts, let me tell you briefly about the National Endowment for the Arts itself.

First, the Endowment is an independent agency of the Federal Government established in 1965. Funding was very small indeed in the early years—some $2.5 million in 1965, climbing to $8.25 million in 1970. These sums were hardly adequate when one realizes that the mandate from Congress was to encourage all the arts—and all the country—in music; dance; theatre; film, radio and television; visual arts and photography; literature; museums; community-based arts; the arts in education; and architecture and environmental arts.

With the leadership of the President and the strong bi-partisan support of the Congress, we have increased the Federal monies in the past four years to some $60.775 million. While this is, perhaps, impressive, the really impressive factor is that all the while private and other governmental monies have been increasing at the same time.

The work of the Endowment is guided by panels of experts and by the National Council on the Arts, an advisory group of twenty-six outstanding artists and leaders appointed by the President.

The design fields are well represented on the Council with the judgment and Texas humor of architect O'Neil Ford, the wisdom and Renaissance character of Charles Eames, and until recently when his term expired, the creativity and insights of landscape architect Larry Halprin.

With the advice of these gentlemen, with the timely interest of the White House, with the encouragement of the Congress, and with the cooperation of the design professions, I believe that a design program is being fashioned at the Endowment that may greatly affect design during America's third century. And, affect it for the better.

Why? Because the program is based on the needs and demands of people; its success will depend on the creativity of individuals and organizations—not on government.

The Endowment's design program is in part carried out through the grant-making process. Last year, the Endowment initiated what is known as "City Edges." We, in effect, stated that as the nation moved toward its third century, the Endowment was prepared to participate on a matching basis with cities or universities or civic organizations that wished to do studies or research aimed at those "edge" conditions in our cities which create both design problems and design opportunities. Freeway edges, river edges, natural terrain edges, edges between old and new sections of towns, even rooftops of buildings (another edge) against the sky—they were all targets. The Endowment received a deluge of worthwhile applications, many more than could be funded; only 37 out of 350 as a matter of fact. The grants ranged from studies of the waterfronts of San Francisco and Portland, Maine, to the uses of New York City's rooftops. One grant helped with urban plans for Lexington, Kentucky's new metro government. Another grant was awarded to Austin, Texas, for a study of waterways to form an integrated open space network for a range of community needs.

And in Philadelphia a grant to the city government will assist the Philadelphia Gateways Committee in a project to improve the effectiveness and environmental quality of such major highways to the city as the Schuylkill Expressway.

The response to the program, and its obvious need, encouraged us to move on during the next few years to assist projects that will enable people to explore those "options" a city might choose that will foster and contribute to the visual qualities that humanize our urban settings.

Cities, like individuals, have choices to make. It is our hope that the new program, called *City Options*—which is descriptive, if not original—will encourage people to really *see* their cities; and to stimulate a new awareness and a new knowledge, both of which are essential if our cities are to be great, if they are to be an art form in our lives. The City Options we seek to encourage are those that deal

2. Effective design of public services is itself an essential public service.
Project: Auditorium Forecourt Fountain. Portland, Oregon.
The Auditorium Forecourt Fountain is an expression of community pride in
Portland. The nation's parks, public buildings, housing projects and trans-
portation systems are the most tangible forms through which Government
reaches the public.
    (Courtesy: National Endowment for the Arts)

with the creation of a cultural and humane environment—one that
takes into account the ultimate purpose of our buildings, their use
and enjoyment by people.

To indicate the nationwide interest that exists in such matters, the
Endowment is receiving almost 300 inquiries a week for application
forms, and the program was only announced in October.

Using the grant process, then, is one way in which the National
Endowment for the Arts can assist people to improve design.

Another equally dramatic effort by the Federal Government had its genesis in May, 1971, when, without much fanfare, President Nixon sent a remarkable memorandum to the heads of Federal departments and agencies. He asked them to direct their attention to two questions: How each agency could most vigorously assist the arts and artists; and second, how the arts and artists could be of help to the agency and its programs. It was the President's strong belief that government should not be merely a patron of the arts, in the traditional sense, but that government should call upon the arts and artists to improve the quality of our society and our lives.

The National Endowment for the Arts was asked to coordinate the replies from the various agencies—and we did. The overwhelming interest expressed by these agencies was to obtain assistance in the design area.

Just one year later, based on the replies of the agencies and the recommendations of the National Council on the Arts, the President issued a message. It was, to the best of my knowledge, the first time since the birth of the nation that design was the subject of a Presidential message. In it, the President said, "There can be no doubt that the Federal Government has an appropriate and critical role to play in encouraging better design." As a beginning, he announced that the government would move ahead on four fronts:

—The Federal Council on the Arts and the Humanities, of which the Endowment's Chairman is a member, would sponsor Design Assemblies for federal administrators and artists.

—The National Endowment for the Arts would appoint a special *ad hoc* task force committee to review and expand the publication, *Guiding Principles for Federal Architecture*, to improve the quality of Federal architecture across the country. The Endowment would also recommend a program for including art works in new Federal buildings.

—The National Endowment for the Arts would also coordinate efforts to upgrade Federal graphics and publications.

—Concurrent with these activities, the Civil Service Commission would review procedures for rating and employing artists for the Federal service.

The Endowment is fully aware of the opportunity represented by design initiative directed by the President. We know that if it is successful in making the Federal Government aware of its responsibilities, that the results stemming from such design awareness could be the basis for enormous impact on all sectors of design in the country's third century. Among other reasons is the fact that one of the worst offenders—and unfortunately one of the largest, most influential clients for design services in this country—is the United States Government.

Since the President's Message, all I can say is "so far, so good."

For example, a Federal Design Assembly was held in April, 1973. Its purpose was to develop a greater awareness among Federal decision-makers—i.e., clients—of the importance of design to their agencies. The Assembly's theme, "The Design Necessity," sought to establish that: 1) effective design of public services is in itself an essential public service; 2) that design is not a luxury or a cosmetic addition; and 3) that good design can save money and time and enhance the effectiveness of Federal programs. The impact of the first Assembly was very positive and very far-ranging. For example, some of the departments are running mini-design assemblies, while some states have expressed keen interest in sponsoring their own state design assemblies as springboards for initiating statewide design improvement programs.

Some of the other ripple effects of the Design Assembly and the President's interest were hard to predict and somewhat surprising to us at first. How would you react to a telephone call from the Federal Prison Industries, for example? Calmly, I would hope. Or how would you like to be advised that two of your staff associates are off to Leavenworth? Equally calmly, I would hope. The reason is that the Federal Prison Industries produce about $60,000,000 worth of products every year—from desks and wastebaskets to filing cabinets and mattresses—which are then purchased by the Defense Department, Veterans Administration and the General Services Administration and used throughout the Federal Government. Prisoners also learn manufacturing and craft skills as they make the products.

In the past, this program has operated with little or no design assistance; and I don't think I have to elaborate concerning the results if you are familiar with government offices and installations.

When Federal Prison Industries representatives came to the Endowment, they asked how they could improve their products, how they could retool and update their plants. Our response was to call upon a private organization, the Industrial Design Society of America. As a result—and as a first step—a network of talent has spread out across the country to visit the various plants and to make evaluations.

Ponder for a minute what it will mean if the efforts toward good design by this *one* Federal agency are successful. Better products, better working environments for government employees, better training for prisoners. This is only one example of the stimulation of ideas created by the Federal Design Assembly.

The second design improvement initiative is directed at the quality of Federal buildings. For the past year a research staff, a distinguished Task Force, and twenty Federal agency representatives have been reviewing the earlier 1962 version of *Guiding Principles for Federal Architecture* to see how the principles might be improved and expanded.

It is the consensus of the Task Force that architect selection procedures and the working relationships of architects with public clients are key concerns. Members have expressed interest in examining innovating public sector models like the New York State University Construction Fund and the Foreign Building Office of the U.S. State Department. The Task Force also has asked the staff to explore new opportunities provided by government space needs, including adaptive use of historic commercial structures, and mixed-uses for government office buildings. During the coming year the Endowment will submit the findings of the study on Federal architecture.

In addition to the Design Assembly and the architecture study, the Endowment took on the seemingly impossible task of improving the design quality of all printed matter produced by the U.S. Government. The objective of this program is to *strengthen visual communications* by raising the design standards of departments and agencies. In addition to establishing close liaison with each of the Federal agencies, the Endowment is enlisting the support and counsel of design professionals outside the government. To date

3. Design enhances communication.
Project: Graphics Program. Internal Revenue Service.
Design is crucial to effective communication. The Internal Revenue Service relies heavily on design to communicate tax information to high school students through these series of brochures.
(Courtesy: National Endowment for the Arts)

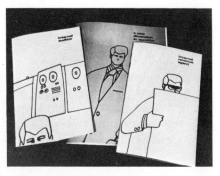

fifteen Federal agencies are taking part in the graphics improvement program, representing twenty-five percent of the total number of Federal agencies. All are having their graphic products reviewed by panels of professional communication and graphic design experts. Following the initial reviews, the panels are reconvened to meet with the agencies' administrators and designers to discuss their evaluations and recommendations for improvement.

Already results are evident. The Labor Department and Civil Service Commission have decided to engage the services of design consultants to propose a comprehensive graphics system for their agencies; the Department of Agriculture has reorganized its graphics services and hired an outstanding designer in a position of leadership. And, the National Zoo—the home of the Chinese pandas—has asked for design assistance so that visitors can find the pandas and other animals!

Without qualified, talented designers and architects in key positions within the government, we cannot hope for success in this venture. Therefore, the fourth part of this program to make our government a better client is one that invests in the people who will be responsible for making design an important part of America's third century. The Civil Service Commission study and the implementation of its recommendations will assure that the Federal Government actively seeks to employ the finest designers and artists in the country and to bring them into the decision process.

The four-part Federal Design Program—with a deadline for significant improvement by 1976—represents an important part of our effort to make the *public* sectors more enlightened design clients and hopefully create a more fertile situation where good design can happen.

On another front, there are some who say that our Federal Government should take a far more active role in encouraging improvement in design in private industry. They note that our position and reputation as a world power has been based on industrial *production* rather than design excellence. The quality of product design has been largely left to the discretion of the individual manufacturers with little government assistance or encouragement. The results of this policy of unintentional neglect by business and by government have been uneven. Some companies, such as IBM, Mobil Oil Co., and Westinghouse, have developed design programs of significance. Most, however, have been content to emulate successful products rather than innovate new products and improve existing ones through sound design research and processes.

In case you might question why any national government would have a role to play in this area, I would like to note that the United States is the only major country in the entire world without a program of active government encouragement for improving industrial design. That fact would appear to signify a rather strange state of affairs—or at least a situation that should be questioned by people in the business community. And, a question which should be posed to the government as well.

Speaking of questions, I am constantly amazed in most discussions of design that the question always arises: "How much will it cost?" That is a perfectly good question to ask, just as long as it is not posed separately from the basic concern itself. Design is not an add-on. It is a matter of doing what you do, but doing it better. It is part of the essence of the thing. The more appropriate question would be: "How much will it cost if we don't incorporate the design factor?" We can see the answer to that question all around us just by looking at how well and efficiently our transportation systems work, how well our products compete on the world market, and how successful we have been in creating truly livable cities.

And, I am equally amazed at the number of people who ask the experts for advice about design at the end of a project. Design

4. Design can save money.
Project: The Acorn School. New York, New York.
The Acorn School used ordinary construction scaffolding to form modular units for classroom use, thus saving a considerable amount of money which would have been spent to furnish the classrooms.
   (Courtesy: National Endowment for the Arts)

questions require expert attention as you begin to ponder a problem—not as an afterthought.

   Too often, as well, consideration as to improving design in this country has been based on the question of what shall we build when we tear the old building down? Fortunately for the nation, we have a significant shift in attitude on that question so that as we look toward the design of the third century, I feel certain that Dulles Airport in Washington will be a booming center for some kind of airborne vehicle well through 2050; that the new Boston City Hall will be the seat of government of that city, still. (In fact, I can

5. Design can save time.
Project: Student Housing. State University College, Brockport, New York.
The fast-track construction system reduces construction time in half.
    (Courtesy: National Endowment for the Arts)

imagine historic preservation groups struggling to save it as a worthy
example of mid-1900's architecture.) And I trust we will have
Monticello—though there seems to be some question at the moment
as to which version we will have! And who knows; the University of
Pennsylvania may again be restoring Houston Hall and still taking
pride in the oldest student union in the country. And, most as-
suredly, people will be coming from all over the country and the
world to view Mount Vernon and to visit the beauty of the well-
designed city of Washington.

Almost everyone knows the history of George Washington, as the
"father of our country," the Commander-in-Chief of the American

forces in the Revolution, and as the first constitutional president of a new government. All are towering achievements, particularly for one man. But even more than these accomplishments, Washington understood that architecture and design were as important for the nation as they were for himself. His private attention was directed toward the planning of Mount Vernon. While most of the architectural elements there were copied by Washington from the English design books of the eighteenth century, the most striking feature of the mansion—its high-columned piazza extending the full length of the house—was his complete innovation which would in itself entitle George Washington to distinction among architects.

But Washington also carried his understanding of the importance of architecture and design into public life as well. His vision for the capital city embodied the hope that a city would emerge from the swamps and wilderness that would be the equal of any European capital and he entrusted to L'Enfant the planning task. His interest in these matters was not motivated by a desire "to keep up" with the Europeans or just to create a beautiful city. Rather, Washington realized that a government, in the nature of its buildings, sets a standard and can have an important influence upon the community or society in which it builds. That building, therefore, should reflect the highest standards toward which that government aspires.

Whether as the head of Tiffany or IBM or a small business; or whether as government official or civic leader, the same understanding of the importance of design should apply. What we build, what we produce, where we live, should reflect the highest standards toward which each of us aspires. And the sum total will be our design for America's third century.

Index

Reference numbers in *italics* indicate illustrations or their captions.

Index

109